No Stone Unturned

A Life Without Bounds

Historical Notes for Context

Donald Andrew Beattie

Apogee Prime

All rights reserved under article two of the Berne Copyright Convention (1971). No part of this book may be reproduced or transmitted in any form or by any means, electronic or mechanical, including photocopying, recording, or by any information storage and retrieval system without permission in writing from the publisher.

Published by Apogee Prime, A division of Griffin Media

No Stone Unturned /Donald A. Beattie
First Edition, Second Printing
© 2011 Apogee Prime/The Author
ISBN 978-1926837-18-5

All photographs are from the author's files unless otherwise noted.
Printed and bound in Canada

Dedication

For Travis and Kyle Beattie and all the Beatties who will follow. As the years go by, I am sure these remembrances will sound more and more like fiction.

Acknowledgments

Many family members and friends made it possible for me to write this autobiography. Some deserve special thanks, as from time to time I sent them excerpts to review and check for accuracy or other special requests.

Harvey Terwilliger added detail to the Hudson River boating adventure. Carol Donovan, Norwood Municipal Clerk, searched the records to find the town map and picture of the school. Jürgen Haffer, before his death, reviewed some of my Colombia stories.

Bob Kowalski was especially helpful critiquing my accounts of flying and tour in VS-30. Wit Johnson, another VS-30 squadron mate, helped me interpret my flight training logbook as he once had a tour as an instructor at Pensacola.

My brother Tom and sons Tom and Bruce provided details of their careers that I have included at the appropriate places.

Three cousins, Margaret Shaw Dohe, Nancy Cleary Patcke and Shirley Brandt Shumar, were kind enough to go through family albums and send pictures. Several are included.

Roger Van Ghent, as he did for two previous books, put the manuscript in a computer-ready form to make it easy for Rob Godwin to publish. Unfortunately, Rob, you almost certainly won't be able to retire on profits from this one or, for that matter, my other book you published.

If you will enjoy reading this story, it is because Mary Beech edited and made it readable and grammatical, no small accomplishment. My Tenafly High School classmate, Mary appears in one of the stories. With great patience she read and reread as I changed a story or added detail. She will be rewarded in this life and the next for her efforts. However, while her back was turned, I would change a sentence or add a few that were never subjected to her critical eye. I take full responsibility for any errors you may find; they are not her fault.

To all the above and many others not mentioned, my thanks and appreciation for responding to my call.

Table of Contents

Introduction   10

Chapter I - Growing up during the Great Depression and WWII   14

Early Memories   14
Norwood, New Jersey   17
Summer Camping   *28*
School Days and Hanging Out   *29*
On to High School   41
Boating on the Hudson   *46*
Applying for NROTC Scholarship   *50*
Brother and Sister Update   *51*

Chapter II - Columbia College 1947 to 1951   63

Freshman Year   63
First Midshipman Cruise - June 1948 to August 1948   *66*
Sophomore Year   73
Second Midshipman Cruise - June 1949 to August 1949   *76*
Junior Year   80
Third Midshipman Cruise - June 1950 to August 1950   *83*
Senior Year   84

Chapter III - Navy Active Duty 1951 to 1952   89

Reporting to USS Liddle   89
USS Liddle - APD-60   *90*
Change of Command   *94*
Applying for Flight Training - Liberty in Naples   *96*
Returning to Little Creek, Virginia   *97*
Operational Readiness Inspection   *98*
Moored in Hampton Roads   *100*
Escorting LSS(L)s Bound for Korea   *101*
Skipper Relieved - Orders for Flight Training   *102*

Chapter IV - Flight Training 1952 to 1953 — 105
Pensacola — 105
Pre-Flight Training — *105*
Whiting Field - Soloing — *107*
Corry Field - Instrument Training and Night Flying — *109*
Saufley Field - Formation Flying and Gunnery Training — *111*
Barin Field Carrier Landing Practice - First Carrier Landings — *112*
Advanced Training — 114
Kingsville, Texas — *114*
Carrier Landings in TBM — *116*
Advanced Instrument Training - Corpus Christi, Texas — *116*

Chapter V - Fleet Squadron VS-30 - 1953 to 1956 — 122
Reporting to VS-30 — 122
Carrier Landings in the AF Guardian — *123*
ASW Training and New Flying Skills — *124*
Night Carrier Landings — *125*
Sixth Fleet Deployment on USS Mindoro — *126*
Norfolk - New Romance — *129*
Operation Blackjack — 130
Transitioning to the S2F Tracker — *133*
Flying the AD Skyraider — *134*
Marriage — *137*
Leaving VS-30 — *138*

Chapter VI - Graduate School - Colombia, S.A. 1956 to 1963 — 143
Colorado School of Mines — 143
Graduate School Studies — *145*
Joining the Navy Ready-Reserve — *146*
Thesis — *147*
Job Offer — *149*
Oil Exploration in Colombia — 151
Field Work Begins — *154*
Exploring the Darién — *163*

Wildcat Wellsite Geologist   *172*
Living in Colombia   *176*
Leaving Colombia   *180*

Chapter VII - Inside the Beltway - 1963 to 1983   195

National Aeronautics and Space Administration 1963 to 1973   195
Science Planning for Apollo and Post-Apollo   *197*
US Geological Survey Joins Our Team   *209*
Using Photography from Earth Orbit   *211*
Searching for Terrestrial Impact Craters   *212*
Developing Apollo Science Experiments   *214*
Commercial Application of Gemini Photos   *216*
Astronaut Application   *217*
Soviet Lunar Space Missions   *222*
Preparing for the First Landing   *223*
Antiwar Demonstrations   *224*
Back to Apollo   *225*
The Flight of Apollo-11   *226*
Family Vacations - Father's Death   *228*
Apollo Missions Continue   *229*
Astronaut Training   *231*
Broadcasting Apollo Missions on Voice of America   *233*
Apollo Epilogue   *235*

National Science Foundation -1973 to 1975   236
Divorce and New Marriage   *238*
New Neighbors   *241*
Energy Research Programs   *243*
Project Independence and Bilateral Energy Agreements   *244*

Energy Research and Development Administration - 1975 to 1977   249
Operating the First Modern Wind Turbine   *251*
1870s Photographs   *252*

Department of Energy - 1977 to 1978   253
Coping as DOE Assistant Secretary   *253*
Leaving DOE   *257*

Returning to NASA - 1978 to 1983   258

Managing Solar Energy Projects	*258*
Three Mile Island Accident - Space Solar Power	*259*
Energy Projects for AID	*261*
New Job Offer	*263*
Colombia Vacation	*263*
Energy Program Terminated	*264*
Tom Begins His Air Force Career	*265*
Bruce Chooses His Future Path	*266*

Chapter VIII - Private Sector Employment - 1983 to 1996 — 280

BDM Vice President — 280
Problems in Houston — *281*
1870s Photographs Revisited — *283*

Return to Maryland - Consulting — 283
Costa Rica - Nicaragua Project — *283*
Back to Colombia — *284*
Vice President George H. W. Bush Election Campaign — *288*
Boeing Space Station Proposal — *289*
NASA Space Station Advisory Committee — *291*
First Grandson Born — *293*
Pakistan — *293*
End to the 1870s Photographs Story — *297*
Mexican Communication Satellite — *297*
Second Grandson Born — *299*
ENDOSAT Unmanned Aerial Vehicle — *299*
History of Solar Energy Research — *302*
Visiting Family and Friends — *303*

Chapter IX - On the Last Lap - 1996 to ??? — 316

Finding a New Home — *316*
First Book — *317*
Getting Organized and Local Politics — *318*
Tom's Participation in War with Yugoslavia — *321*
Second Book — *322*
Tom Retires from Air Force - Begins Career as SW Airlines Pilot — *324*
Terrorist Attacks of 9/11/01 — *325*
Mother's Move from Florida — *326*

VS-30 Reunions   *327*
Superbowl XXXIX   *328*
Third Book   *329*

Afterword   331
Final Thoughts - Pondering the Imponderable   *334*

Index   344

References   350

Introduction

Early in the morning of Wednesday, October 30, 1929, I was born on Staten Island, New York, the second son of James Francis and Evelyn Margaret Beattie, nee Hickey. The next day, Halloween, my mother celebrated her twenty-seventh birthday in bed holding a seven-pound six-ounce trick or treat. The day before I was born, on what would come to be called "Black Tuesday," the stock market crashed, signaling the beginning of the Great Depression. Three important events occurred in three days and I don't remember any of them, amazing! Will future historians connect the Stock Market crash to my pending birth? Those who follow Chaos theory and the "butterfly effect" may find a connection. It couldn't have been my fault, could it?

My brother, Thomas Anthony, had been born three years earlier in Chicago. My sister, Margaret Ann, would be born in June 1932, also on Staten Island. My perception is we weren't poor and we weren't rich. Perhaps the most important advantage we had as a family was that my father had a steady job throughout the Depression.

My mother, born in Chicago in 1902, was the oldest of six children; she had two brothers and three sisters. Her father, whom we called Pa, was a Chicago trolley motorman. Her mother Muz was a housewife, surely very busy tending to such a large family. After graduating from high school my mother started work at the *Chicago Daily News* reporting to the composing room superintendent, Andrew Adair. He called her his "gal Friday," meaning she did whatever was needed to help him get the paper out on time. After her marriage and she became pregnant with my brother, she left her job. Adair told her she would be welcomed back whenever she wanted to return. She never did, and the family moved to Staten Island, New York, in 1928. Muz and Pa visited occasionally, so they were familiar grandparents. Aunts, uncles and cousins on my mother's side visited occasionally in the summer. You will read about my mother's many skills in the stories that follow.

My father, born in Chicago in April 1901, was the youngest of four, one girl and three boys, the oldest being a stepbrother. I never knew much about his mother and father; at least I don't recall their being mentioned often around the dining room table. His sister Margaret and brother Edward with their families also visited. I don't recall any visits from his brother John. My father's early life is discussed in some detail as you read on.

With that brief introduction I begin this long tale, probably of interest

Brooklyn Daily Eagle front page, Wednesday, October 30, 1929.

Evelyn Margaret Hickey, age five.

only to the author or, perhaps, relatives and brave readers researching earlier times. As I sit at the keyboard pecking away with one finger, wishing I had taken typing in school, jumping from one story to another and different eras as the mood dictates, I find memories flooding in of faces, names and events I would have thought were long forgotten and irretrievable. Often I return to a story, adding details that surface unexpectedly from the murky depths. One's mind is apparently a wonderful museum only waiting for a visit. Are all past events and scenes indelibly saved somewhere on its dusty shelves? Sometimes, after spending a few hours during the day writing, I awake in the middle of the night remembering a past adventure; 3:00 AM seems to be an especially productive time. Beside my bed I keep a pad, jot down a few words to remember the story in the morning, and go back to sleep.

As one grows older, recalling earlier events and describing them to others is an enjoyable way to spend an hour, hopefully not repeating them too often to the same audience. In my case I have missed telling some of the stories that follow to wide-eyed grandchildren sitting at my feet and protesting, "Grandpa, you are making it up that you didn't watch television when you were a little boy." Also, my grandchildren have not heard the usual tale, "When I was your age, in the winter I walked a mile to school in deep snow." This of course I did, just a half-mile. But sometimes the snow was pretty deep.

My grandchildren are on their own and never had to suffer through such sessions. Perhaps when their lives slow down they and their children, assuming they have families, will have time to read about family history and this life story. I have tried to capture what it was like growing up in the 1930s and '40s, a period of great change not only in America but throughout the world. The years that followed included events we would have thought possible only in science fiction stories when we were children. Jet planes with hundreds of passengers crossed the Atlantic Ocean in a few hours, astronauts walked on the Moon, spacecraft landed on Mars (whatever happened to the little red men?), or a telephone carried in one's pocket showed a picture of the person you were talking to. Amazing!

What follows are everyday stories from a long life, 29,000 days and counting, interspersed at times with highlights I hope you will find interesting. Some are enhanced with photos collected at the end of each chapter. As I flipped through old copies of the *National Geographic Magazine* as a very young boy it was the pictures, especially those of bare-breasted African maidens, that caught my eye. Sorry, none of that type will be found in the next pages. Unfortunately, photos couldn't be found of most of the events and people I describe in the 1930s and '40s.

Cameras weren't as available then as in later years. And if photos were taken they weren't always saved. I hope those I have included will be helpful in visualizing ancient times.

To spare you boring details or because of concerns for family privacy, I have been discriminating in the stories I have included. I believe the early stories reflect how the Great Depression and growing up in a small town influenced my life. That decade, immediately followed by WW II, had a lasting effect, instilling the traits of thrift, hard work, perseverance and assuming responsibility for one's actions. Also, throughout my adult years I have followed old Satchel Paige's wise advice: "Don't look back, something may be gaining on you." I visited or worked in all 50 states and some 40 countries. You will find a few stories that took place in what were, at the time, out of the way locations. However, the world seems to be a much smaller place today than when I first opened my eyes on an October morning in a Staten Island hospital.

Father with siblings, John, Margaret and Edward (1906?).

Mother and Tom, Chicago, 1928.

Chapter I
Growing up during the Great Depression and WWII

Early Memories

One of my first recollections is playing with our dog, a collie-German shepherd mix named Lassie. She was a wonderful, friendly pet but would bark loudly at strangers, making my mother feel secure when home alone. Undoubtedly Lassie's gentle nature and affection for all of us created my love for dogs. A small boy can't experience any greater friendship than lying in the cool grass next to the family dog, gently stroking its silky-soft ears, or a big boy for that matter. I know; I once was both. As an adult, when possible, I have had a dog.

The first home I remember was three stories high and out in the country. In the 1930s rural areas were common on Staten Island, the least populated of New York City's five boroughs. Our house was rented from, I believe, the Benning family. I don't recall if there were any nearby homes or neighborhood children. The house backed up to a wooded area, and I played by myself in the woods. I dimly remember sitting on the roof over the front porch in the afternoon, contrary to my mother's wishes, watching the road that ran in front of the house and waiting for my father to come home from work.

We soon moved to a two-floor (plus basement) bungalow rented, I believe, from the Bendell family. This home was in a more settled area, with many others nearby. Stairs led up to the porch and front door, and a detached garage and tool shed were in the back. The lot was deeper than it was wide. Lassie was allowed to run free in the unfenced yard.

In the living room-parlor was a baby grand piano that my mother played. I recall sitting next to her on the piano bench and singing from a Raggedy Ann's Sunny Songs book. I still have the tattered copy and remember the words and tunes of some of the songs. Care to sing along?

"Little Wooden Willy, people thought him silly

'cause he had a knothole in the middle of his head.

So he put a hat on, and since he had that on,

folks who called him silly think he's very wise instead."

I wonder what my mother was thinking as she encouraged my interest in music and singing. Will this little guy go on to perform great deeds after these first steps? Or, in a few minutes I will have to get up and do the laundry. I'll never know because I never asked her such a deep question. Sons don't; perhaps daughters are not as inhibited and have those conversations. One might expect it was the former thought; if so, I hope I lived up to her expectations. Mothers everywhere are optimists. If they weren't, the human race would have disappeared long ago.

We had an icebox in the kitchen, and a man would deliver a large piece of ice every few days. He would chip off a piece from a large block in the back of his truck to fit in the top of the icebox and charge by the pound. On hot days we would grab the chips to cool our mouths. Milk was delivered to the front door in glass bottles and left in a wooden box where we put the empties for the milkman to pick up. There was a paper seal on the bottles. Cream would separate at the top. When we wanted a glass we would shake the bottle to mix the cream with the rest of the milk, carefully holding the cap down.

The same procedure was used for another favorite food, peanut butter. Peanut oil would be separated at the top of the jar and had to be stirred with a spoon to mix with the thicker paste. Even after stirring the peanut butter was still thick and stuck to the roof of one's mouth. My father would eat it by the spoonful, a custom I never enjoyed. However, the homogenized variety is still a favorite food. The process of homogenizing foods, starting with milk, began in 1919. Apparently by the early '30s few in the food industry had adopted the process because, I suspect, of initial equipment cost.

In this new neighborhood we played with children who lived a few doors away. One of our favorite places was under a neighbor's front porch. It became a kind of club house, too low for adults to crawl into but just the right height for kids. We didn't have a swing set or other forms of backyard entertainment common today, but we played hide and seek, cowboys and Indians, and stayed in the neighborhood.

I don't ever remember being bored; there was always something to do. One favorite pastime for my brother and me was to set up small lead soldier figures that our father made. He poured melted lead into molds of figures and types of military equipment. When it was cool we would remove the realistic pieces, trim off any excess lead, and paint them. We then had small armies to maneuver in the house or back yard and pretended to be famous generals. We had a few wind-up toys; remotely controlled vehicles hadn't been invented yet.

My mother, who preferred to be called Margaret although my father always called her Evie, was a very good cook. My father had a few specialties and always made the gravy from the roasts. He was a sailor in WW I and among other duties had been a cook and carpenter. After Navy service he attended the Illinois Institute of Technology. I don't believe he pursued a formal engineering degree, instead going to work for the Bell Telephone Company with a subsequent transfer to its subsidiary Western Electric in New York City. He never dwelled on his engineering accomplishments, but he worked on advancing teletype technology resulting in at least one patent that was held by Western Electric. He commuted daily to Manhattan, taking the Staten Island ferry for five cents each way. Then he took the subway to his office, also for a nickel.

An event I still remember quite vividly happened one night, perhaps when I was three or four. We heard an airplane fly low over the house. This was very unusual as there were not many airplanes flying in 1933, especially at night. When we heard one during the day if we weren't outside we would run out to watch it. The closest airports were located across the harbor on Long Island. The next morning a neighbor came by and said a plane had crashed a few miles away. We drove to the crash site, and there was the plane (possibly a Ford Trimotor). I remember looking inside and observing that the seats were made of woven wicker like a garden or porch chair. I think everyone survived, as I have a dim picture in my mind of the plane not seriously damaged. However, when I called the *Staten Island Advance* newspaper to get more details, its archives had no record of the crash. Perhaps this is research for another day.

Several other memories of special times before moving to New Jersey in 1935 filter up out of the past. My father was active in the American Legion. The local Staten Island chapter held annual picnics. It was only 17 years since the end of WW I, and there were many active veterans, so the Fourth of July picnics were well attended. There were lots to eat and games to play. We sat at long tables listening to the veterans reliving their service. The big treat was an ice cream cone topped with a Borden's Mello-Roll vanilla bar, the only flavor available, that came wrapped in paper before it was peeled off and placed on the cone.

In April 1917, the month the US entered WW I, my father turned 16. He tried to enlist in the Navy by lying about his age. According to his story, he didn't qualify as he was underweight. He went to a store and bought and ate some bananas. Then he went to a different recruiter and passed. Apparently a birth certificate wasn't required. He served on two ships during and shortly after WW I, the USS Chattanooga, a light cruiser, and the USS Beaufort. The Chattanooga saw mostly convoy escort duty during the war. The Beaufort, a seized German coaling ship, served in the European theater. Some U.S. Navy ships in those days were still fueled by coal, thus the need for coal supply ships. I recall only one story my father told about liberty while on the Beaufort at Port-au-Prince, Haiti. He and some shipmates rented donkeys to ride around the island, eating bananas taken from the trees. He still liked bananas.

My brother and I attended a Catholic school run by French nuns. Only three events related to kindergarten remain after all these years. I remember walking to school with my brother after a heavy snowstorm left shoveled snow as high as my head. We were taught some French and sang French songs. I remember only one:
Frere Jacques, frere Jacques,

Dormez vous, dormez vous.

I think that's how it went; in any case those are the only words I remember after 75 years. The nuns carried a ruler and would slap us on the wrist if we weren't paying attention. The fact that I remember this must indicate that I was slapped a few times, although hitting a kindergartener seems a little extreme.

I spent only a few months in kindergarten. My attendance was shortened as I contracted mastoiditis and was very sick. I was taken out of school and didn't complete the year. Antibiotics didn't exist, so the treatment was just to let the illness run its course. I remember my mother placing me on a cot outside in the sun and swabbing my throat frequently with a dark, foul-tasting solution called Argyrol. However, a common treatment for children with that infection was to make an incision behind the ear and drain the cavity. I recovered and didn't need that operation. The next year when we moved to Norwood, New Jersey, I was able to enroll in first grade with my age group. I recall sitting behind a few boys of my age in school, including my new friend Charlie Ford, and seeing the ugly dark scars that remained - long, narrow cavities behind their ears. I don't recall seeing any girls with that operation. Perhaps it was hidden behind their long hair.

Norwood, New Jersey

It was rumored that the large house my parents bought in Norwood was built in colonial times and was over 200 years old. We later learned that it was built in 1836, so the rumors were off by only 100 years. It had been a farm house and across the street was almost a twin. Both had been built by the Haring family, the original owners who had bought 16,000 acres from the Tappan Indians in 1636. I believe our property was a little less than one acre. The house across the street was owned by the Frese family, a brother and three sisters. (Our house was written up in the *Old-House Interiors* spring 1998 issue after later owners restored it to what was considered close to its original condition.)

Between the two houses was Tappan Road, the major north-south artery. To the north it went through Northvale and then passed one of George Washington's Revolutionary War headquarters in Tappan just across the New York border. To the south it ran through the next town, Harrington Park, and then connected to other towns and roads leading to the George Washington Bridge and New York City. Almost all roads at this time were just two lanes.

Another rumor had it that our house was the first in Norwood to have indoor plumbing. Who knows? The one bathroom was very small, just large enough to accommodate a tub, toilet and small washstand. Our plumbing was connected to a cesspool that my father enlarged by adding a drain field. A man with a big tank truck came every year or so to pump out the system.

The house was built in three sections. I assume each addition was added as the family grew in size and wealth. The first section was the smallest with a very low attic used for storage. Before we moved in this section had been converted to a one-car garage. The west wall was replaced with garage doors. The owners (those written about in *Old House Interiors*) made it into a kitchen and replaced the garage doors with a large fireplace as it was in the original house.

The next larger section was a full two stories with the dining room and kitchen on the first floor, my bedroom on the second floor, plus a small room for my mother's foot-treadle Singer sewing machine. When my mother's parents visited in Norwood Pa slept on a twin bed in my bedroom snoring loudly throughout the night; a pillow pulled over my head to soften the din. This part of the house was lower than the next section, one step lower on the first floor, two steps on the second floor. My father built a fake fireplace in the dining room where once there had been a functional one. The arched stone base that supported the original fireplace was still in the basement. Behind the fireplace a narrow stairway led up to my bedroom and the sewing room. The entrance from outside was a Dutch two-piece door on the south side.

The third section was much larger than the adjoining section with a small front porch and large door bordered by narrow windows. The first floor windows were large with twelve panes and wooden shutters; the second floor windows were small and narrow, typical of the design of homes of this era. This section consisted of a main hall and two interconnected living rooms (parlors), both with fireplaces. A stairway led up to a wide landing, where a large cedar chest was placed, three bedrooms and the bathroom. A low attic was above the second floor.

Under the second and third sections, in a dirt-floored basement, were a gas hot water heater and furnace. Many of the main beams still had their original bark. The basement was dimly lit, and spiders spun webs and hid in the beams. One of my jobs was to fill the furnace at night and during the day take out the ashes and dump them at the back of the property. Because of his asthma my brother was excused from such chores.

We used coke as fuel, as it burned more evenly and would last all night when the furnace grate was filled. Coke was delivered by horse-drawn wagon rather than truck because entry into the lot on the north side of the house was difficult. The driver, the same old man every time, would shovel two tons down a chute into a large bin my father built next to the furnace.

My father did a lot of remodeling. In addition to the fireplace in the dining room he built wainscoting and chair rail around the walls and a corner cabinet. While remodeling the dining room we all worked at removing layers of wallpaper that had been added over the years, a long tedious job. When we reached the bare plaster my father covered the walls with a scenic wallpaper in my mother's favorite color, best described as robin's egg blue. He separated the two parlors in the main section, removed a door on the side facing the road, and built bookcases and an aquarium in the north parlor.

After two years in Norwood my sister and I had our tonsils and adenoids removed. Doctor Johnson, our family physician, said they were the cause of my frequent colds and earaches. He operated on the dining room table. My father administered the anesthesia, and we recovered in my sister's bedroom. Our recovery was aided by a lot of ice cream, a real treat in those days, because it soothed our sore throats. Doctor Johnson was right; I have had few colds and no earaches since. However, I had one bad reaction to the operation. For many years afterward I had recurring nightmares of being suffocated and would wake up in terror. I believe they were linked to the anesthesia my father administered.

As a seven-year-old I was also haunted by thoughts of incapacitating sickness and death. My friend Charlie Ford had suffered a terrible accident while playing and lost his right eye. He now had a glass one but technology of the day was not very good and it was very noticeable. My brother's best friend Tommy Edwards was diagnosed with bone cancer. He lost his right leg, but the disease spread and he died at age twelve. Chemotherapy hadn't been discovered or he might have survived. But my biggest fear was of contracting polio. It would be years before the Salk vaccine put those fears to rest. It was quite common to see pictures of children in iron lungs, or with crippled and disfigured arms and legs. How would I react if such a terrible disease attacked me? It was a scary thought for a very active seven-year-old. But I escaped the dreaded disease, and none of my friends were ever afflicted.

When we moved to Norwood my father refinished the piano, and it stood in the south parlor for the next 20 years. I took piano lessons for a short time; my sister stuck with it and became an accomplished pianist. A short digression to complete this story: when my parents returned in the late 1950s from a tour in Germany where my father worked for the U.S. Army Signal Corps, the piano had been in storage, was in poor shape, and was sold.

My father also re-shingled the roof with the help of a friend. He built a chicken coop at the back of the property where we raised 20 to 30 chickens during the war. That was another chore, shoveling out chicken droppings every week or so. It was not a pleasant job; among other odors the droppings generated ammonia gas. On the property were three pear trees, two apple trees, a black cherry and quince, walnut trees, black raspberry bushes and a blackberry patch. In the spring the many lilac bushes filled the air with their lovely, almost overpowering fragrance. In the summer my sister and I sometimes slept in the backyard in a homemade tent under a large sycamore tree.

A quarter-acre vegetable garden was on the north side of our property. We grew cauliflower, corn, broccoli, Brussels sprouts, tomatoes, string beans, radishes and onions. The third year after we moved to Norwood my father planted two long rows of asparagus. This was a difficult job, and helped by a friend they dug trenches about two feet deep to set the asparagus roots. Two years later we had wonderful fresh asparagus all summer. Strawberries were planted between the asparagus rows. My father built a wire fence around three sides of the garden to keep out rabbits, of which there were many, and a wooden picket fence on the south side facing the backyard.

Every spring my father would get a Burpee's seed catalog and order what he wanted for the garden. He built low greenhouses with glass lids along the south wall of the garage. When the seeds arrived he put them in shallow dirt trays in the greenhouses until they sprouted and were large enough to transplant to the garden. During the war this was called a "victory garden," but we had one long before the war started. As kids our job was to weed and help gather the harvest. My mother canned and preserved the vegetables and fruit. We would wrap the apples in newspaper, put them in a large crock, and store them in the cool basement for winter eating.

At the height of the Depression my mother fed the homeless men (called tramps) who would appear occasionally walking along Tappan Road. She was sure that they had made a secret mark on our house to tell those who came along that they could get a meal if they asked politely. I don't remember if she asked them to do any work in return; perhaps she did.

There was a large open field to the south, formerly farm land, with a few very large apple trees in which we built houses. My sister and I dug a foxhole in the middle of the field and roofed it over with boards. The field was overgrown with fox grass so our foxhole was hidden. A few years after we moved to Norwood the Amicuccis built a house on the hill at the south end of the field. To the west there were no houses all the way to the New York Central West Shore Railroad and far beyond. When discussing with unbelieving citified college friends how rural Norwood was in the 1930s, I told them it was possible to walk all the way to Canada from our backyard without seeing any signs of civilization except when crossing Route 59, the only east-west road near the border between New Jersey and New York and then farther north the Erie Canal. Of course I never made that trip; my boast was based on knowing the immediate area and studying maps.

Rivervale Avenue, a dirt road, ran east to west down the north side of our property ending at the railroad tracks. On the other side of the road was a large property with six small cottages built by a dentist who lived in New York City. He would come with his extended family to spend the summers. He had built a large, raised concrete swimming pool, very unusual in those days for someone's backyard, and a handball court. He must have had a very good practice. In the winter after they left the pool was always drained, and we used it and the handball court backboard to practice hitting tennis balls with my parents' old wooden-framed racquets. There was also one large older unoccupied brick home that was probably the original farm house for the property. The house had three cellar levels that we were told were used to raise mushrooms. It was scary going down in the cellars, like being in a cave. There were no lights, so we didn't do that very often. The house was heavily damaged by fire a few years after we arrived. The property was then bought by an evangelical minister who held revival meetings during the summer characterized by loud singing and praying. My brother and I had a few disagreements with the two brothers who lived there, but my sister became a close friend of their sister, Grace Anzevino.

Across the street, south of the Freses' house, was another large open field. It was used once a year by the town fire department, all volunteers, for their fund-raising carnival. Rides included a ferris wheel, merry-go-round, and all sorts of tents with games and activities. It was noisy but fun. For one week we could look out our bedroom windows at night and watch all the goings on. The rest of the year the field was empty and I would fly my kites, attached to a very long string roll my father supplied, blown far to the east by the prevailing winds.

Beyond the open field on the east side of Tappan Road were three more houses, two very large. The wealthiest man in town, Henry Essig, lived in one with his sister Minnie. He had been the town's first mayor. The second large house was vacant when we arrived; perhaps the Depression had forced the owner to leave. The smaller house belonged to long-time residents, the Jacobs family. On the hill behind their house was a tiny log cabin in which an old couple lived. The woman, Mrs. View, was Mrs. Jacobs' mother. Mrs. View was our baby-sitter when needed.

Kite flying became a lesson in basic aerodynamics, although at the time I wasn't aware of the connection. I built versions using pieces salvaged from old kites and newspaper and wood scraps from my father's work bench. I experimented with the proper design and length of the tail and the curvature of the cross beam that stretched the paper. However, my homemade kites never flew as well as the store-bought ones. Later, as a pilot, somewhat knowledgeable about the history of aviation, I found it interesting that the Wright brothers flew kites at Kill Devil Hills, North Carolina, before attempting powered flight. Movies that show them running alongside their models like little boys, as they flew off the sand dunes, reminded me of my kite flying days.

Since many families did not have phones in their homes, the fire department was called by hammering on a gong made from a railroad rail curled into a big "C" hung from a wooden frame. A large hammer was chained to the frame. It was definitely frowned upon to ring the gong as a prank. Each gong, placed about every quarter mile on the side of a main road, had a coded number for its location. The code was the number of times and cadence one hit the gong with the hammer. There was a gong across the street at the corner of Tappan Road and Hudson Avenue.

If someone rang a gong, it would be heard at the police station a half-mile or so away or by someone who would alert the officer on duty next door to the fire station. He would then call the volunteers, who would go to the fire station to drive the two trucks or go directly to the gong location, knowing where it was based by how it was rung. The only tavern in town, Hasenstab's, was across the street from the firehouse, so usually some men were quickly available to man the trucks. This source of volunteers led to a few jokes but no serious problems.

We had a telephone. Picking up the earpiece would turn on a light at the main switchboard in Closter, and a woman operator would ask, "Number please?" So few phones were hooked up that our first phone number was just two or three digits. I never used a phone at a friend's house to call home. By the time I went to high school many more homes and businesses had phones; the numbers were longer and started with a two-letter prefix.

The northern part of town was an Italian enclave. Many of my classmates with family names such as Benaquista, Scaglione and Bocchino had grandparents living with them who had been born in Italy. They came to the US in the early 1900s, passing through Ellis Island like millions before them. They maintained their old country culture, many making wine from grape vines and the common chokecherry tree. For me this ethnic diversity provided a wonderful advantage, as I would occasionally be invited to have Sunday dinner with my friend Bobby Sticco's family. It was a bountiful feast of Italian dishes, homemade wine, and lively conversation among three generations sitting at the table. The next town to the north, Northvale, was almost entirely settled by Italian immigrants and their offspring.

Norwood was divided into two parts, Norwood and West Norwood, although West Norwood existed only as a post office designation. We had mail boxes at both post offices, why I don't know, but after every school day my job was to collect mail from both boxes. They were a little less than a mile apart, so in addition to other outdoor activities I got lots of exercise just collecting the mail. West Norwood was serviced by the New York Central, with the post office in the railroad station. The other post office was serviced by the Erie Railroad on which my father commuted to New York City. The "Weary Erie" eventually went out of business and my father commuted on a bus that ran by our house on Tappan Road.

All the trains were pulled by steam locomotives with the largest called a 4-8-4 (four front wheels, eight drive wheels, four following wheels) running on the New York Central. To show how brave we were we would sit at the side of the tracks, put a penny on the rail and watch it flatten as the locomotive wheels rolled over it. With a little luck we would find it when it was flipped off the rail. We didn't do this very often as a penny was big bucks. A few pennies could buy important things such as candy or chewing gum with baseball cards attached.

About 75 miles north was the U.S. Military Academy at West Point, also located on the West Shore railroad. My father and some friends often attended the Army-Navy football game, usually held in Philadelphia, and rooted for Navy. The Army team would travel by train to the game. Once, when we learned the time they would pass through West Norwood, my brother and I stood by the tracks at the end of Rivervale Avenue with a sign that read "Sink the Army." Of course we never knew if the team saw it. In spite of this show of disapproval we occasionally visited West Point for special events; watching the cadets marching and maneuvering in perfect unison was a thrilling sight. I was impressed with West Point and its beautiful campus and thought about attending it one day. A favorite book, *Herb Kent - West Point Cadet*, told about life at West Point, glossing over, I learned later, the difficult challenges of a cadet's closely regimented life.

We didn't have a washer or dryer. My mother would wash the clothes in a deep double sink in the small laundry room attached to the kitchen. For the tough stains she had a washboard and would rub the garment over the corrugated surface. Sons guaranteed that mothers would find tough stains. Then the clothes would be hung outside on a line to dry. In winter they were hung in the laundry room to keep them from freezing although this room was unheated and sometimes when taken down they would be stiff. My father's shirts were sent to a commercial laundry.

My mother would iron the clothes in the north parlor-library. When I was five or six I remember lying on the floor while she ironed, listening to Mary Margaret McBride on the radio. She had a lovely, soothing voice, told great stories, read letters from listeners and unobtrusively mixed in words from her sponsors. Sometimes my mother listened to a program that featured an organist and canaries. When the organ began to play the canaries would sing loudly. I can't imagine that such a program would have a big following today, unless perhaps on an iPod if they are still in use when you read this.

Behind our house was a low hill that I concluded was a glacial moraine. A small quarry had been dug in the side of the hill to remove gravel for roads but was now abandoned. Mixed with the gravel was a wide variety of larger rocks pushed ahead of the glacier that covered the region some 20,000 years earlier. As it moved south, over hundreds of miles before stopping in our back yard, it collected material along the way. I would spend hours splitting the rocks to see what minerals were inside. I built a wooden compartmented box to keep my collection and on appropriate occasions showed it at school. I knew the names of only a few of the minerals but was impressed with their variety and color. Thus began my interest in geology.

Also at this time I received a microscope for Christmas. I always wanted a chemistry set but never got one, perhaps showing the wisdom of my parents. I would examine the rocks under the microscope. I also started a bug collection. I put out pie pans of water at night and the next day collected the dead or dying bugs that crawled into the pans and couldn't escape. I saw hairy legs, compound eyes, filmy wings and other interesting bug anatomy all revealed for the first time. Based on my careful study of the human race and activities that boys enjoy, it is clear that rocks, bugs and boys were meant for each other.

Christmas was a very special time until I was about six years old and finally realized that everything was done by my parents. I don't recall specific Christmases on Staten Island but I do remember those in Norwood. At first our Christmas tree was bought and hidden across the street at the Freses' house. On Christmas Eve, after we were asleep, my father would bring it into the house. He and my mother and neighbors would decorate it and put the presents under it. When we awoke, as if by magic, there was the tree in the south parlor. Once we understood how it all came to pass the tree was set up and decorated by all of us before Christmas eve, but presents under the tree on Christmas morning were always surprises.

My father must have been one of the first to decorate with miniature lights. All my friends' trees had large ones about the size of today's candelabra lights, as they were the only type one could buy. But my father wired together tiny telephone switchboard lights that he got at Western Electric and dyed them different colors. We had the only tree with these unique, tiny lights. My mother took great pride in showing off the tree with its hundreds of lights to the neighbors. She also took special care in hanging the tinsel. It was bought as a heavy consolidated mass of thin shreds of lead that I haven't seen for many years. She carefully separated each shred from the mass and placed it so that it hung straight down from its branch and would look like an icicle. The tree was also decorated with ornaments, some very elaborate old German blown glass of various designs. Our trees were beautiful, and Mother would continue to hang or rearrange strands of tinsel almost to the day we took the tree down.

Our town library was a converted house across the street from the school. My favorite early readings were Tom Swift and Hardy Boys adventures. Although the Tom Swift stories were set in the early nineteen hundreds his many inventions, high performance airplanes, submarines, big guns and other unusual devices were exciting to read about.

Then I discovered mystery stories and went through the library's total collection of Ellery Queen, S. S. Van Dine and others. I don't recall reading Agatha Christie mysteries. I can remember the distinctive smell of the library, perhaps from the binding and glue that held the pages in the books. We were allowed to browse the collections housed on two floors. I often went to the library after school.

One Christmas I received Mark Twain's *The Adventures of Tom Sawyer*. Before the day was out I had read it twice. No one writes stories like that anymore. Other favorite books were the "Jerry Todd" and "Poppy Ott" series written under the pseudonym Leo Edwards, exciting and funny boy adventures set in the 1920s and '30s. Eventually over the years we had the whole series, about 20 books. Few were kept as my mother and father moved to new homes. I later purchased as many as I could find in used book stores. They were read by sons Tom and Bruce and grandsons.

Two other books had a great influence on my view of the grown-up world, *King Arthur and the Knights of the Round Table* and *Myths and Legends of Greece and Rome*. Both were nicely illustrated and told of heroic deeds in times past, some of which I took as true - King Arthur, the only one of pure heart able to extract Excalibur from the stone, brave knights of the Round Table going on long pilgrimages to vanquish evil knights and rescue damsels in distress. With my homemade wooden sword I would venture out in the nearby fields slaying evil, chopping down arrogant and ugly bushes. Take that, Black Knight! And Ulysses' adventures all seemed possible in an ancient, mostly unexplored world. The Trojan War was another example of rescuing a fair maiden stolen from her home. All were adventures a boy could dream of imitating.

I didn't pay much attention to my clothes. In the winter I usually wore corduroy knee pants that made a zipping sound when I walked, especially when they were new; high socks; shoes that occasionally had holes in the soles covered by a cardboard insert; shirts and sweaters. I learned to darn my socks, which seemed always to have holes in the heels and toes. The corduroy pants would also wear out and eventually have holes in the knees. At that time it wasn't fashionable to wear pants with holes in them. If they couldn't be patched and were beyond repair, they were finally thrown in the garbage can.

Our winter coats featured mittens attached to the sleeves with elastic bands so we wouldn't lose them, at least very often. In the summer we wore sneakers until our toes went through the ends or the soles fell off. Remember, this was the height of the Depression. Clothing was handed down from brother to brother, not thrown away until it was completely useless. As the younger brother I got the hand-me-downs that still had life in them. But I grew faster and taller than my brother and eventually escaped this ritual.

Every year before Easter my mother would take us to Gimbels in New York City to get new suits and other clothes. For some reason she never shopped at Macy's. She would go to the same salesman in Gimbels' clothing department, who knew her by name. When my brother went to high school in 1939 he often wore still serviceable suits sent by my Uncle James in Chicago. I never wore a suit to high school, nor did any of my friends.

There were three churches in town - Catholic, Presbyterian and Episcopalian. The Catholic church was in a former two-story clubhouse building. The Presbyterians worshipped in a small white wooden church with a tall pointed bell tower just two blocks from our home. The Episcopalian church was a lovely Tudor-style design made of field stone with a long arched walkway from the pastor's home to the church. Almost everyone went to church on Sunday, and churches were full on Christmas and Easter. On Ash Wednesday Catholic and Episcopalian kids attended school with black smudges on their foreheads, but that was the only distinction one might notice between faiths.

Dressing up for church on Sunday was not a favorite activity. We attended the Immaculate Conception Catholic Church across the street from school. We walked to school and church, about a half mile from home. The priests were from the Carmelite order. My mother liked the first priest, Father Dominic, but was not too fond of his successor, Father Charles. My brother and I were altar boys in grade school and often served at daily mass before school as well as on Sundays. I always had a problem remembering when to ring the bells. We all made our First Communion and Confirmation at this church. For the latter, in my picture I wore a dark blazer and white pants, probably the same ones my brother wore in his picture three years earlier. It is one of the few photos showing me wearing a coat and tie until my college days.

Going to confession for my First Communion at age seven was strange and stressful and remained so in later years. The confessional was at the back of the church, a three-part wooden enclosure. The priest sat in the middle behind a closed door and on either side were dark curtained-off compartments in which one knelt. Click - a little sliding door opened and I could dimly see the priest. I knew he was there because I could smell cigarette smoke and sweat. "Bless me father for I have sinned; it has been two weeks since my last confession (different time periods between confessions calculated for each occasion after the First Communion)." Then a few venial sins were confessed (never a mortal sin) with some slightly exaggerated to make the effort worthwhile. Based on the seriousness of the sins an appropriate number of Hail Marys and Our Fathers were levied, occasionally the longer Apostles' Creed if a sin deserved more punishment. They were quickly and silently recited in a nearby pew. After that I had a clean slate and was free to sin again, an early introduction to the world of risk, reward and punishment.

My father's interest in gardening and flowers led to a close friendship with the owner of a local nursery, Ludwig Luck. He came to the US from Germany before WW I. He met his wife in the US and together they established a nursery on 100 acres along the eastern border of town. They had two children, son Ludwig Jr., my brother's age, and daughter Marie, my age. We visited often, and while my father and Mr. Luck walked around the property my brother and Ludwig went off and played together. My sister and I played with Marie. My mother visited with Mrs. Luck, probably exchanging recipes. At times we stayed for dinner.

The Lucks lived in an old Dutch colonial house with thick stone walls and fireplaces. Around the property were a number of outbuildings. We would climb on the tractors and other machinery, roam the carefully cultivated fields and explore the buildings. Marie, a pretty blonde, was my first romance. But at six or seven it could definitely be categorized as "puppy love." I don't recall that it was reciprocated. Except for my sister, girls were already a mystery. Marie married, has a son and daughter, and now lives in South Jersey. I haven't seen her in many years, but we communicate occasionally by letter and phone.

Summer Camping

When we moved to Norwood we had a 1932 black, four-door Dodge with a spare tire mounted on each front fender. We made one trip to Chicago in this car; I believe it took three days each way. At night we stayed in small cabins, the motels of the day.

We also drove to Waterbury, Connecticut, in the summer to visit my father's friend Bill Lucien, with whom he had served in WW I. It was a trip, according to the odometer, of exactly 100 miles. Starting in mid-morning it took the better part of the day to reach Waterbury via 9W, the Bear Mountain Bridge, and narrow two-lane roads in Connecticut. One way we amused ourselves (are we there yet?) was reading the Burma Shave comical rhymes painted on barns or signs along the road. We competed as to who would first see a sign and call out the complete rhyme as it was painted on sequential signs along the road.

"Just spread then pat"... *"More shaves that's that"*... *"Burma Shave."*

"Hot tip pal"... *"More shaves per gal"*... *"Burma Shave."*

(I don't claim to have remembered them; jingles are courtesy of Google)

We would arrive in mid-afternoon and before leaving for the campsite have a big late lunch featuring Italian dishes made by members of Bill's extended family. Every summer Bill set up a camp at Black Rock State Park, and we would stay for two weeks. The camp consisted of two large umbrella tents plus a cooking and eating area with a dining table, two Coleman stoves, and an icebox. A large tarp covered all. A few other families camped nearby, and we knew some of them after so many visits. We brought most of our food as there were no stores nearby. There was no electricity, just Coleman lanterns and flashlights; however, ice was delivered for those with ice boxes.

Nearby was a small lake where we fished, without much success, and many trails running up into the mountain. It was an idyllic place where we could roam free and pretend to have exciting adventures. I can still remember how the forest smelled in the morning, a pleasant, musky aroma of damp earth, pine needles, ferns and wild flowers. High on the mountain were a few small water-filled rock gorges where we could swim. The most exciting activity was swinging on a long rope suspended from a tall tree that carried us far out over the edge of a sheer cliff while sitting on a thick stick, the closest thing to flying a small boy could experience. As I swung out off the cliff the pit of my stomach felt that it was about to drop to my knees - great fun!

School Days and Hanging Out

In Norwood I started in first grade. Our teacher was Mrs. Agnes Mc-Neely, who lived just a few blocks south of the school at the corner of Summit Street and Blanche Avenue. My most vivid memories of first grade are of enjoying drawing and singing in the mixed choir for the Christmas pageant. Jimmy Newman, who could have been my twin, and I led the procession as the choir filed into the auditorium. I recall being told not to sing so loud.

Norwood's public school was typical of the time. The main building, part brick, part wood, the latter painted red, had two stories plus a basement and attic. There was a small tower where the bell was rung in the morning and after lunch to call us to class. The first floor of the main building held four classrooms, kindergarten through third grade, plus the principal's office. On the second floor were three classrooms, sixth grade through eighth, plus a small auditorium. The auditorium had a raised stage with a curtain, and the audience sat on folding chairs. The attic, with a pull-down stair, was used for storage. In the basement were a large furnace (hot air heat, no air-conditioning) and supplies for the janitor. Behind the main building on the east side was an enclosed walkway of some 50 feet that led to a T-shaped, one-story extension with classrooms for grades four and five. The total student body was about 250.

In front of all the classrooms were blackboards, a clock, an American flag and a painting of George Washington. We said the Pledge of Allegiance and recited the Lord's Prayer each morning, the latter coming with two endings. Those of the Catholic faith ended with "and deliver us from evil, amen." Those of the Protestant faiths continued with "For thine is the kingdom and the power and the glory, forever, amen." The two different endings never seemed to bother anyone. If students were of the Jewish faith or atheists, I assume they just remained silent. No protests to this ritual ever surfaced and I didn't pay any attention as to who belonged to which faith.

Usually there was a mid-morning break when we were given graham crackers, a container of milk and sometimes an apple. I never questioned who provided this food; at that age and in that era we accepted what our elders and teachers told us. There must have been a government program associated with the Depression that made our snack possible.

At the end of the day some would stay and help the teacher tidy up the room for the next day. The blackboards were cleaned with a wet sponge and the felt erasers taken outside, clapped together to remove the chalk dust, and returned. Each class had a homeroom mother, sometimes two, who organized special events and provided refreshments. My mother often volunteered for this duty.

On the north side was a playground for boys, on the south a playground for girls. Never the twain should meet - a wise separation as we boys played a variety of rough games before school started in the morning, after returning from lunch, and during daily recesses. Girls did such exciting things as jumping rope and playing dodge ball, and there were swings on their side. Kids in kindergarten and first grade stayed on the girls' side. We didn't have a cafeteria. The few students who lived in Yorkview brought their lunch; the rest of us went home to eat, making the round trip in one hour. Each day Marie Luck was driven to school by one of her father's workers, taken home for lunch, returned, and picked up again at the end of the day.

One favorite game was "knife." Most boys carried a small pocket knife. The game started by marking out a large rectangle in the dirt (there was very little grass in the play area, worn bare through the years by the feet of many small boys), and dividing it in half. The two opponents would stand in their respective halves and take turns throwing their knives into the ground of their opponent's half. If the knife stuck upright, then a line would be drawn, according to the direction the blade faced, and that piece of the opponent's ground was added to the knife thrower's territory. One continued throwing and whittling down his opponent's territory. If the knife didn't stick it was the opponent's turn to throw his knife and try to recapture lost territory. Usually several of these games would be going on at the same time.

As the game progressed, based on the skill of the knife thrower, one opponent's territory would be slowly reduced. To stay in the game one had to be able to stand in his territory to throw his knife. Standing on one foot was allowed if there was enough territory. Eventually the loser would end up in a piece too small to stand in. Sound complicated? It wasn't, and the games could be quickly decided. There were never any stabbings or accidents that I recall. I have gone to this great length describing a simple child's game to show how society has dramatically changed in a relatively short time. Today, boys wouldn't be allowed to play this game at school.

Second grade started serious studies. The teacher was Mrs. Mabel Carter. Math was a little more complicated, with addition, subtraction, multiplication and division, and we diagrammed sentences into their various grammatical parts. As I was a good reader, Mrs. Carter would have me sit with other classmates to sound out words. I don't recall any science studies, but it was in this grade that I started my rock collection. Grades three, four and five don't register in my memory as especially noteworthy.

However, in third grade I managed to pull off a very embarrassing caper. At first I wasn't going to write about it, but an autobiography should include events one would like to remain hidden as well as the many great triumphs readers might find difficult to believe. One Friday afternoon, while playing at my friend Barry Sweeney's house, time passed quickly and Mrs. Sweeney received a call from my mother reminding me that it was almost six o'clock. Oh-oh, my father would be home before I could arrive and there would be at least a lecture. Spare the rod was not practiced in the 1930s; I don't remember being spanked although I am sure it happened. Such punishment seems to be erased from my memory.

What does an eight-year-old do when confronted with such a dilemma? He makes a snap decision and follows through. I decided that, rather than go home to face punishment, I would walk four miles by back country roads to Westwood. Why Westwood? Because every Friday evening my parents would load us in the car and drive us there to buy the next week's groceries. I visualized arriving before them and, to their great surprise, walking up and saying hello. It never occurred to me that they would be concerned that I hadn't returned home. After all, I knew all the time where I was.

Well, they were more than concerned. When I didn't show up for dinner, they called the police and a search was started. In the meantime, after waiting in vain at the store, tired and hungry, I decided to walk back home. My careful plan was in total disarray. It was dark, probably about nine o'clock; every time I saw a car's lights approaching I moved far away from the road. I made it back about three of the four miles when some searchers caught sight of me and I was rescued - captured actually, as I didn't know the searchers and was very reluctant to go with them. Returned to my parents, I was given an unsmiling hug. No corporal punishment was administered, but the local paper had a headline and story the next day. Monday, returning to school, I faced many questions from schoolmates and made up a silly story that I just wanted to see if I could walk to Westwood.

Although television had been an experimental technology for some years, it was not available to the general public. Among other reasons a set was very expensive and none of the major networks were broadcasting programs. However, radio had a big place in our lives. Comic book ads at that time featured a kit using a "cat's whisker" to build one's own radio. I always wanted to buy a kit and assemble it, but the price was beyond my budget.

Nonetheless, about this time I was given a small Philco vacuum tube radio, about 6x4x10 inches in size. These are important dimensions because at night I would hide it under my blanket to listen to a program. It had a lighted dial so I could select a station. I emphasize vacuum tube as that was the technology of the day, and those who read this will probably never have seen such a radio. All electronics used vacuum tubes, and when a tube burned out it could be replaced. One had to ask for the tube with its unique number and letter code as each tube served a different function.

The family radio, in the north sitting room, was much larger and housed in a free-standing cabinet. Besides the evening news we tuned almost religiously to Jack Benny, Fred Allen, Fibber McGee and Molly, Amos and Andy, the Lone Ranger, and Lux Radio Theater. Many were broadcast on Sunday nights and we sat together listening, kids on the floor.

My favorites were broadcast on weekday afternoons from five to six: Mandrake the Magician, the Shadow, Jack Armstrong, and Little Orphan Annie, each in 15-minute segments. Usually, if I was playing ball or some other activity, I would hurry home to listen, as they were all serials. If I missed a broadcast I was out of date on the plot.

Jack Armstrong's sponsor was Wheaties cereal:
"Jack Armstrong never tires of them, and neither will you, so just buy Wheaties, the best breakfast food in the land!"

I didn't like Wheaties; it became a soggy mess as soon as milk was poured on it. To avoid an argument at breakfast my mother stopped buying Wheaties, but I never missed an episode.

Little Orphan Annie listeners had an added incentive to catch every program. At the end of some programs a coded message would be sent that only those who had a decoding device could understand. I received my decoder by sending in the top of an Ovaltine container and 25 cents. Every so often the decoder was discovered by the bad guys and listeners had to send for a new one if they wanted to continue decoding messages. Other devices offered from time to time included a ring that had a small mirror on top to see if anyone was following, or a tiny compass. One ring glowed in the dark; I learned it had a little radium in it. Undoubtedly that finger received a radioactive dose of some size, but I still have all ten. No boy could resist such intrigue; at least I couldn't. We would entreat our mother to buy a can of Ovaltine to mix in our milk when we heard that some exciting offer was available. Most of the time we drank our milk without Ovaltine when we didn't need any more tops.

At six o'clock Uncle Don came on, telling stories, giving advice and, at the end of each program, telling children whose parents wrote in where to find their birthday presents. He was an early version of Mr. Rodgers without the visuals. After many years of broadcasting, always in a cheerful voice, Uncle Don had a comeuppance. Thinking his microphone was off, he was heard to say, "That should hold the little (censored) for awhile." His program was canceled.

I also listened to broadcasts of the Brooklyn Dodgers baseball games and kept a notebook of game statistics using my own statistical shorthand. Red Barber was the reporter who kept my attention by describing all the different ways a player swung at a pitch, fielded a grounder, or delivered a pitch. Born in Mississippi, he had a southern speech pattern and soft drawl. At times he would interject unusual colloquialisms such as "tearing up the pea patch" or "sitting in the catbird seat" or "eating high on the hog" to describe the action.

Barber's knowledge of the physical and playing characteristics of players on all the teams was especially important when describing an away game. In the background the teletype machine clattered (sponsors couldn't afford to pay for onsite reporting of away games), and he kept my interest by filling in the action based on his vast knowledge. Night games were one of the reasons I hid my radio under the covers, especially when the Dodgers were away playing the Cubs or Cardinals whose home games began one time zone later.

A favorite hobby was sitting at the large tilt-top desk my father made for me, building model airplanes and listening to radio programs. Authentic-looking flying-model airplane kits could be bought for 25 cents. The parts were cut out with a razor from thin, preprinted balsa wood sheets and glued together. The models were constructed almost like a real airplane in a factory. The fuselage and wings were assembled piece by piece - main spar in the wings, ribs to give them an aerodynamic shape, fuselage bulkheads with stringers along the sides. After finishing the fuselage I installed the motor, a long rubber band attached to the propellor by a thin metal hook that passed through a small bead in the nose to reduce friction. Then I attached the loose end of the rubber band around a pin in the tail. The finished product was covered with a thin paper skin shrunk tight by gently spraying it with water and painted. The final touch was gluing on the insignia. All of this was accomplished using a tube of glue I bought but small boys aren't allowed to purchase by themselves today.

I flew the plane in the house and outdoors until, after a few hard landings, it was ready for its final flight. Opening the upstairs bedroom window facing Tappan Road, I cranked up the propellor, lit a match to the tail, and sent it spiraling down to a spectacular crash and burn. How realistic can one get? Then it was time to build another plane as soon as I had saved a quarter.

Another Christmas present was a camera with a film developing and printing kit. I made the large closet in my bedroom into a dark room, complete with red light. Using the kit, I developed and printed pictures myself. The kit came with a small tank to develop the film. In the dark room I wound the film in the tank and added the chemicals that came with it. Then I rinsed and fixed the developed film, dried it, and projected it through an enlarger onto print paper. These latter operations would be done with the red light on. The result, after another chemical bath, was a print. I knew no one would accidentally interrupt and spoil a print by opening the door and letting in light because I had rigged the door with a crude make-or-break switch and a buzzer that went off if it was opened, all powered by large dry-cell batteries my father gave me.

One afternoon, in the summer of 1938, I observed some construction on the eastern horizon, where the top of the Palisades met the sky. Work continued for some time, and eventually a very tall tower was built, easily visible to the naked eye. With a little research I learned it was built by Edwin Armstrong, inventor of FM radio. It was the second tower to transmit commercial FM programs; the first in Boston had started transmitting a month and a half earlier in May 1939. Because FM (frequency modulation) transmissions are at a higher frequency than AM transmissions they are not distorted by interference from thunderstorms or other sources of static. With its great height and location in Alpine at the top of the Palisades the tower enabled programs from W2XMN, the station's call sign, to be heard throughout the metropolitan area and as far away as Connecticut. In 1938 very few radios could pick up FM, including ours, but its inherent advantage was obvious and soon FM became a standard broadcast band included on most radios sold.

September 2, 1938, was a memorable day. A severe hurricane, nameless in those days, struck without warning. The previous day had been overcast with heavy rain falling off and on. When the high winds finally hit, some estimated to gust over 70 miles an hour, the rain-soaked soil could not support trees and power poles. Large trees in our neighborhood were uprooted, and the next morning on Tappan Road many of the poles were leaning at crazy angles. I don't recall that we lost power, but

I suspect we did. As the hurricane swept northeastward it devastated areas along the shores of Long Island, Connecticut, Rhode Island and farther north, claiming some 700 lives. Now, during hurricane season, a documentary of the hurricane of 1938 is often shown on television with pictures of the widespread damage. However, there was a positive side to the disaster. It served as a wake-up call to the Weather Bureau, showing the need to better understand hurricanes and to develop the ability to predict their course and strength. Eventually this led to the development of computer models and satellites that improved our ability to track hurricanes as soon as they form off West Africa.

Just before seventh grade began in 1941 the Terwilligers, with two boys and two girls, moved to Norwood. In so doing they increased Norwood's population by four-tenths of one percent (1940 census 1512). During the Depression families did not move frequently as they do today, so this was an important development. The Terwilligers' home was on a low hill overlooking the West Norwood post office, a passenger stop on the New York Central railroad. Mr. Terwilliger worked for the New York Central. The home was a short distance off the path I took each day to pick up the mail. One of the boys, Harvey, became a classmate and we quickly found we had many common interests. Together we shared many childhood adventures, and we remained friends until his death in 2011.

Although Harvey and I had no serious disagreements through the years that I recall, we agreed to disagree on a very important issue, the city with the best baseball team. Harvey was a rabid Chicago Cubs fan; I was devoted to the Brooklyn Dodgers. The two teams were in the National League made up of only eight teams at the time, so they played each other 20 times during the season. We collected baseball cards and kept up to date on all the important team statistics, meanwhile arguing over who had the best pitchers and hitters.

In the summer of 1941 the Cubs were the better team, but there was always "wait till next year" when the Dodgers would prevail. Major league baseball was the king of professional sports; football and basketball didn't command the attention they now receive. In the off-season the "hot stove league" dominated sports pages as writers tried to predict next season's results and keep fans in a state of turmoil. Eventually the Dodgers improved and won a World Series while the Cubs declined. As of this writing, most likely because of the curse of the goat, the Cubs have not returned to the World Series since 1908. Touché, Harvey!

One interest we shared was fishing. The Hackensack River was the main waterway in northeastern New Jersey and several tributaries ran close to Norwood. We would ride our bikes (mine originally my brother's) to the closest stream and spend the day fishing, usually without success. To pursue our mutual interests such as fishing required cash to buy needed equipment and supplies, but jobs for young boys were few and far between. In the summer of 1942 we finally found work cutting the lawns of Henry Essig Jr. and Harold Schoonmaker.

Mr. Essig, an elderly gentleman, lived with his sister Minnie on Tappan Road just a half block from my home. Their house was an impressive white Victorian with a large detached garage, all surrounded by a chain-link fence. A curved gravel driveway circled the front yard. In the center of the front lawn on a pedestal was an exact small-scale replica of the house. I believe Mr Essig was a retired automotive executive. Way back in 1905 he had been Norwood's first mayor. He had a large black car, and the town police chief Tommy Sposa and patrolman Otto Lang were his chauffeurs when off duty. He had a lawn service of some sort before Harvey and I offered our own expert services.

Mr. Essig was a fastidious, frugal man. Probably that's why he chose us over his lawn service because he paid us each 25 cents an hour. Or perhaps he just wanted to reward two eager young boys. In any case, we worked that summer under close supervision. He would often come out to critique our work of cutting the lawn with push mowers, weeding the flower gardens, and raking the gravel driveway. If he thought something we did wasn't up to his standards he told us so, demonstrated how he wanted it done, and had us redo whatever was needed. Once a week we would work on his property for about two hours. Fifty cents could buy a lot in 1942.

We also worked for a short time for Mr. Schoonmaker, who had a large home on Summit Street south of the school. After a snow storm we went door to door offering to shovel snow but didn't get many customers. Times were tight and our neighbors did their own shoveling.

Winters were cold. Although it snowed off and on from November through March, I don't remember many heavy snow falls. However, one year after a heavy snow, on the hill behind our house, my brother made a ski track that ran to the creek at the bottom of the hill. He had been given a pair of skis and boots at Christmas and became quite proficient. We used his ski run for our sleds and ended up on the frozen creek. Temperatures were cold enough to keep the creek frozen for most of the winter. I received a pair of hockey skates and stick and with my friends skated down to the old mill pond in Harrington Park for pickup hockey games. Sometimes we had to shovel snow off the pond to play.

When I was eleven the war in Europe was in the headlines every day and the possibility that the US might soon be involved became a major preoccupation for me. Stories told of German submarines sinking merchant ships within sight of the New Jersey coast. On the evening news we heard Edward R. Murrow's sonorous voice describing the effects of the previous night's bombing of London and other cities. There wasn't much good news; France had fallen and England stood alone against Hitler and Mussolini. Although President Roosevelt had declared we were a neutral country, Lend-Lease had begun and he had transferred 50 WW I destroyers to England to protect the convoys carrying supplies from the US. We subscribed to the local daily *Bergen Evening Record,* and my father brought home two papers every night, the *Daily News* that he read on the subway on the way to work and the *Journal American* that he read on the way home. I began keeping a scrapbook of the maps showing the advances of the Axis in Europe and Africa.

Our position of neutrality changed on Sunday, December 7, 1941, when Japan attacked Pearl Harbor in the Hawaiian Islands. Like many others we were listening to the New York Giants football game on the radio when the broadcast was interrupted with the news. This prompted a quick search on the library shelves for the atlas. Two days later newspapers had pictures of huge black clouds rising above sinking ships of the Pacific fleet and stories of the many casualties and damage. When President Roosevelt declared war against Japan on December 8 and three days later against Germany and Italy, our lives changed dramatically.

Rationing began for almost everything we purchased - gas, clothing, shoes, meat, sugar, butter and more. Because no one in the family needed to drive to any war effort work, we were classified as nonessential gasoline users and the Ration Board assigned us to the A category. My father's job was considered essential because Western Electric was making critical communication equipment for the Army and Navy, but he could commute using public transportation. My brother, 15 at the time, had two years of high school remaining before he became eligible for the draft.

The A category entitled us to four gallons of gas each week. Since our 1937 Oldsmobile straight-eight engine provided only about twelve miles per gallon, long trips were out of the question. There would be no more summer camping in Connecticut. We carefully economized. My Mother kept the ration coupons for all the essentials. When she ordered meat from Loidl's butcher shop, trimmed to her very specific instructions, I took the coupons and gave them to Mr. Loidl. Sugar rationing had the greatest impact, especially in the fall when my mother preserved fruit and vegetables and made several kinds of jellies. Despite rationing, some items were not always available. But we learned how to manage and considered ourselves fortunate. We understood how difficult it was for those in England to survive the shortages they endured, not to mention the constant bombing. Our vegetable garden, chickens, fruit trees and bushes all contributed. We were never hungry.

As we weren't addicted to watching television or using computers, we passed the time with simple pleasures. On fall afternoons we often played touch football on Adam Street that ran between the homes of Jimmy Frosco and Johnny Kohlberger. It was an unpaved cinder road, and play always resulted in bruised knees and skin coated with grey dust. After these games I ran home, about a half mile. I established a routine - run as fast as I could between two power poles, then walk to the next one. I ran and walked all the way, and by the time I got to high school I was in pretty good shape. If the weather didn't permit outside play we adjourned to the pool table in Johnny's front living room. Through the years we wore a path in the rug around the table, much to his mother's displeasure. Some nights we played cards alternating the host house. Some evenings, especially during the summer, long walks around town helped us burn off a little energy.

Also at age eleven boys begin to understand the world around them and what it takes to survive in an adult world. With this knowledge comes the need to assert independence. Interests expand as one's abilities grow. Today it means starting in middle school and, among other discoveries, noticing that girls might be fun to be with if only to tease. In Norwood it meant moving to the second floor of the school that contained grades six through eight, leaving the "little kids" on the first floor. In grades six to eight we had homerooms of about 30 students and rotated teachers for the first time, each of the three teachers specializing in one or more subjects.

Every month the school published a newspaper, *The Teller*. It was the creative work of grades six through eight under the supervision of seventh grade teacher Mrs. Hubner. Every edition contained at least one story from each of the nine grades plus artistic contributions. Printing was done on a mimeograph machine with stencils typed by students. We belonged to the National Junior Scholastic Press Association and submitted copies for judging. Each year a convention was held in New York City and a few of us attended. Going to the "big" city, representing the school and comparing our newspaper with newspapers from other schools was a big deal. It was an even bigger deal when we brought back prizes, which we did.

Norwood had a Boy Scout troop for many years, but it was disbanded in 1936 before I was old enough to join. A new troop was formed in the spring of 1943, and I joined. With help from my parents and some money saved I bought the official scout uniform, including the wide-brimmed state trooper-style hat. On meeting days or special occasions we wore our uniforms to school to impress our female classmates. I quickly passed the tenderfoot requirements and then those of second class scout.

Norwood Troop 120 energetically supported the war effort. We conducted scrap metal and waste paper drives, scouring the town for whatever we could find. It took a little courage, but wearing our uniforms we went door-to-door asking residents to buy defense stamps and war bonds. Defense stamps were sold mostly in ten, 25 and 50-cent denominations. Savers would paste the ten and 25-cent stamps in a book until they had $18.75, at which point it could be exchanged at the post office for a bond worth $25 at maturity ten years later. If you were saving ten cent stamps, the last nickel was paid in cash. Or, if buying 50-cent stamps, you saved until you had $37.50 and received a bond worth $50 dollars at maturity. I don't recall how successful we were, but my largest individual sale was a $37.50 bond bought by the rector of the Episcopalian church.

I began working on the tests to become a first class scout. One required passing a difficult swimming test that included saving another person in deep water. I wasn't a swimmer at the time and there were no nearby facilities at which to learn. I couldn't see how I could pass that test within a reasonable period. So, despite being an active participant in Troop 120's early and important war activities, I remained in scouting less than a year. None of my close friends had joined, and starting high school that September involved many new activities.

Our teachers prepared us well for the next step, high school, but Norwood did not automatically promote every student to the next grade; promotion had to be earned. Studies hadn't been "dumbed down" to permit all students to pass. As a result, by the time I reached eighth grade two classmates, both boys, were two years older. Among other bad habits they smoked outside class. In class they sat at the back of the room and rarely participated.

Miss Burns was our eighth grade homeroom teacher and taught Social Science. She was probably in her mid-thirties (although I wasn't a good judge of a woman's age) and somewhat attractive. She had a habit while lecturing or sitting at her desk of putting her hand inside her blouse and rubbing her breast. That habit caught the eyes of all the boys, especially the older ones sitting in the back. Outside class we would talk about it. I always wondered if it was an unintentional gesture or one to keep our attention. If so, it worked but may have been a distraction from the subject being taught. I wonder if the girls in the class observed this and thought it strange. I'll never know, but the two older guys are listed in the graduation program; they finally graduated despite this distraction.

My final day as an eighth grader was supposed to be an exciting one. However, I wasn't at school to participate. I was at home with a case of chicken pox. As required, a lady from the health department had come by and pasted a yellow sign on the door stating that the house was under quarantine. In anticipation of the final day I had been selected to write the class prophecy. My mother attended in my place and did the reading that was supposed to be both humorous and prophetic. I will never know if it caused any laughs.

I wrote the prophecy as if I were a radio announcer recalling events that took place in 1955. I reported accomplishments of 26 noted classmates who had quickly gained fame in many fields (really quickly in just four years after college, but I had great classmates). They included splitting the atom, sending rocket ships to Venus, and other fantastic events. I reported that my friend Harvey pitched no-hitters in both games of a double header for the Chicago Cubs and was nominated for the Baseball Hall of Fame.

Predicting the atom would be split wasn't bad for June 1943. The ongoing Manhattan Project at Los Alamos was a close-held secret unknown to a thirteen-year-old. One prediction, that Richie Vogler would be mayor of the large city of Norwood, famous for its medical institutions and companies run by classmates, almost came true. Years later he was elected to the town council, served for 17 years, and then became the town administrator. I could tell he was a born politician. One part of the prophecy didn't come true; Norwood is still a small town.

I found the two-page typed attempt at deathless, prophetic prose tucked away with my diploma and graduation program. With only two Cs all year, in art class, I am the only boy listed among five honor students. This appears to be an early example of sex discrimination against other male classmates who I knew had pretty good grades.

My mother performed one other important function. Norwood's graduates attended high school in either Closter or Tenafly, with each school receiving equal numbers of graduates. Closter was primarily a vocational high school emphasizing classes to prepare graduates for work in various trades. Tenafly was primarily a college prep school providing courses in higher math, foreign languages and science, classes not offered at Closter. Tenafly also offered vocational classes. This year more of us wanted to attend Tenafly than Closter. To resolve the problem lots were drawn from a jar.

The drawing was done alphabetically by last name. Bernie Baudish was first but because he wasn't planning on going to high school he didn't have to draw. Thus my mother was the first, and she drew Tenafly. My closest friends, Harvey Terwilliger and Bobby Sticco, also drew Tenafly. Two other friends, Jimmy Frosco and Bobby Daw, wanted to go to Tenafly but were unlucky and pulled Closter lots. Jimmy then decided to go to St. Cecilia High School in Englewood; as a result, his parents had to pay tuition at the private school.

On to High School

And so I began the next chapter in my life, certainly one of the most important. Attending Tenafly High School shaped all that came after. To this day friendships made in high school continue, and the Class of 1947 has met regularly every five years since graduation.

Tenafly is located about five miles south of Norwood. In the early 1940s it was an upper middle class commuter town of some 8,000 with quiet shady streets, many large homes, and a small business district that included a movie theater. It was a stop on the Erie Railroad and was serviced by two bus companies. The high school received students from Alpine, Harrington Park, Demarest, Haworth, and Cresskill as well as Norwood. Most of the student body was from Tenafly and had known each other through elementary school. The small group of students from Norwood, scattered through four grades, were considered by some of the more urbane Tenafly students as hicks from the country as we commuted by bus from the "wilds" of far northern New Jersey. However, during my time there we competed well academically, and many of the varsity athletes came from the sending towns.

By the time I started high school I was earning pocket money by setting pins at a bowling alley in Closter. Johnny Kohlberger, Harvey and I worked there several nights a week, usually from about seven to eleven at night, and caught the last bus home. The tenpin bowling alley had manual pin setting rather than automatic. After each ball was bowled we jumped down into the back of the alley where the downed pins collected, picked up the ball and placed it on the return track. Then we carefully cleared any pins still on the alley. After the second ball, if the first wasn't a strike, we stepped on a lever that raised steel spikes in the shape of a V to reset all ten, picked up the downed pins which had a hole in the bottom, stood them over the steel spikes, and released the lever. We stacked the pins four at a time, two in each hand, then jumped on the thick leather bumper at the end of the alley. Sometimes a pin would ricochet up and hit us, but we soon learned which bowlers were throwing a fast ball and took precautions. If we wanted to get a tip, we did all this as fast as possible.

If we were really fast, as I became, we could do two adjacent alleys simultaneously. We got ten cents a line, 40 cents per game, $1.20 for a three-game set. If there were other bowlers waiting we could work another set, and by working two alleys take home five bucks. Not bad for four hours of work, but we earned it. On nights when there was league play we were very busy, and the winning teams gave us a nice tip. I spent my money on lunches at the Tenafly Diner, comic books, and other essentials.

A short, sad story about my comic books: I began reading and collecting them about 1938 whenever I had a spare dime. My favorites were Superman, Batman, Spiderman, Dick Tracy, Flash Gordon, Green Lantern and a few others. I stored them in the bottom drawer of the dresser in my bedroom. From time to time I would take one out and reread it, then carefully place it back in the drawer. The bottom drawer was the largest in the dresser, and through the years it held dozens of comic books in near pristine condition.

When I went away to college my comic book reading days were left far behind. My collection, I thought, was safely in the bottom drawer of my dresser back home in Norwood, although at the time comic book collecting was not a big hobby and it really wasn't a concern. However, a few years later my parents moved and my mother, probably like thousands of mothers before her, decided these cheap little magazines didn't deserve to be moved, and they were trashed. There went probably a few million dollars in the comic book market. However, if mothers kept all their sons' comic books the market would look different today. It is a delightful fantasy to think what I might have sold them for if I had the foresight to keep them for the next 30 years, but perhaps my wife would have thrown them away when I was on a trip.

Often on nights when we weren't working Harvey, Johnny and I would hitchhike to Dumont or Bergenfield and back. Almost always friendly drivers would pick us up, but sometimes we had a long walk home. During the war hitchhiking was a common way to get around, and it never occurred to us that there might be danger in accepting a ride.

Attending Tenafly High School was like stepping into a new world. From the time I caught the bus in the morning until late in the afternoon I was on my own with old and new friends. At the start of my freshman year we students received some notoriety by staging a strike in support of football coach Leonard Kachel, who had been dismissed by the school board without any reason given. As a new freshman I had never had any contact with him but, based on what older students told me, it seemed the right thing to do. We left the high school and marched to the Commons a few blocks away where the football field and track were located and milled around for a few hours. New York papers took note and wrote about our protest, which lasted three days. However, our strike had no effect as the reason for Kachel's dismissal was never revealed.

Actually, I wasn't a freshman in high school. I was in ninth grade of junior high school taking classes in the high school building. In June 1944 I received a Certificate of Admission to the senior high school, having satisfactorily completed my ninth grade classes. School work was challenging and enjoyable. Our teachers were wonderful professionals dedicated to advancing our knowledge. When our class meets at reunions we tell over and over the same stories about our favorites and their idiosyncrasies that made attending class enjoyable.

During the four years I took all the math and science offered and two foreign languages, Spanish and German. Our German teacher, Madame Dubensky, who also taught French, was a refugee from the Soviet Union, having fled to Germany during the Russian revolution and then come to the US. A tiny woman, she was strict and brooked no nonsense in class. As a result she was not always a favorite, but I got along fine with her and in my second year of German was selected president of the German Club.

In addition to the college prep courses, at my father's urging I took a vocational course, mechanical drawing. It was taught by another war refugee, Mr. Motyl, who demanded perfection in all we did. We drew precise cutaway renderings of machinery and parts as well as architectural subjects. This class really helped me to visualize the internal structure of different types of objects and to read blueprints. When I went to graduate school many years later I was glad to have had this background. I highly recommend such a course to any aspiring engineer.

Along with class work I tried out for football and track in my sophomore year. My attempt to make the football team was short-lived. I weighed only 130 pounds, and I guess Coach Yockers couldn't see me making much of a contribution although I could run faster than most of the other aspirants. Harvey and Bob Sticco made the team, and at 155 pounds played guard; 200-pound football players were a rarity. Most of the good athletes played both offense and defense. I don't remember that the school had a weight room or even free weights. There have been enormous changes in the support of high school athletics since the 1940s.

Tenafly played in a very competitive North Jersey league. We won our share of games against much larger high schools in Rutherford and Englewood. Perhaps the most interesting rivalry was the first game of the year against St. Cecilia, the private school Jimmy Frosco attended.

It could recruit and enroll student-athletes from any town and at one point had won over 30 consecutive football games. The coach of the team was Vince Lombardi, a relative unknown at the time, who went on to coach at Army, the Giants and the Green Bay Packers. He also taught chemistry and physics and Jimmy Frosco said he was a favorite teacher. We never beat St. Cecilia while I was at Tenafly. Lombardi ended his coaching career with the Washington Redskins during the time I had two season tickets.

I made the track team and ran in several events over the next three years, including the 440, 880 and mile relay. My times were pretty good, and I won a number of events competing against other high schools. Harvey, Bob and Johnny were also on the team. Eventually we all earned varsity letters. My black wool sweater, with an orange letter "T" on the front, moth hole free, is still in a drawer.

Recently, paging through my high school yearbook the *Tenakin*, I noticed a demographic of the Class of '47 which had never occurred to me while in school. Of the 137 in the class only three were minorities, one Asian, one with a Middle Eastern surname, and one African-American. The first was a girl, the latter two boys. Willy Hatten, the African-American, came from Cresskill and was a good athlete. He was a running back on the football team and during track season ran the 100 and 220 and at times the mile relay with me. He also boxed in the "Golden Gloves" and one year made it to the semifinals. We were good friends throughout the four years. When he learned I would be going to Columbia he wrote in my yearbook that I should go out for the track team, which I did.

My social life with members of the opposite sex was not very extensive; I was too involved in more important activities. It was fun to be around the girls and I had a few close friends of the opposite sex, but dating was not a high priority. Based on our mutual admiration of the Brooklyn Dodgers Mary Beech was one of those friends. In our junior year we cut afternoon classes and took the bus and the BMT subway to attend opening day at Ebbets Field. When we arrived the game had been rained out. We got back on the subway and went to a movie at the Paramount Theater in Manhattan. Not to be denied, the next day we cut classes again and saw Ed Head pitch a no-hitter. Sometimes second chances work out beyond expectations. Mary and I still stay in touch. She is retired after teaching learning-handicapped teenagers in California. If you read the Acknowledgements, you saw her name there and the important role she played in this story.

Although my varied interests didn't emphasize dating, at 16 I got my driver's license and drove our 1937 Oldsmobile on weekends. As I recall, only two guys had their own cars and drove them to school. Having a car available in my senior year with gas rationing ended allowed me to date a very pretty girl, a junior who lived in Haworth. Her name will be a secret to protect the innocent from questions that might be raised by the millions who will read this book. That was as close to a romance as I experienced in high school and was certainly an enjoyable relationship while it lasted. Now married, she lives in Florida but I have not seen her since 1947.

In the summer of 1945, between my sophomore and junior years at Tenafly, I managed to land a job at Camp Shanks, located about a mile and a half north of the New Jersey/New York border. Camp Shanks had been one of the major east coast embarkation facilities during WW II for troops being sent to Europe. To pick up soldiers troop ships sailed up the Hudson River, passing under the George Washington Bridge. A few miles farther north they moored at the Piermont Pier in the Tappan Zee, the widest part of the river.

Victory in Europe was declared on May 8, 1945 (VE Day). Harvey, Johnny and I boarded a bus to Manhattan, then took the subway to Times Square to join the celebration. Soon troops were returning from Europe in preparation for being sent to the Pacific for the invasion of Japan. Camp Shanks became a debarkation destination for these soldiers, housing them in wooden barracks until trains were available to take them to the west coast. In June it was a very busy place, and I convinced the "management" that, although a few months shy of my sixteenth birthday, I should be hired. Male workers were still in short supply. It was my first full-time job with a paycheck. Every day I rode my bike the two plus miles to the gate and then to my job restocking the PX, a camp store. PX stock included every item imaginable from men's and women's clothing to candy bars. For the returning soldiers, many away for a year or more, it was their first opportunity to buy things they hadn't seen since they left the States. A favorite purchase was nylon or silk stockings, items that wives or girl friends had difficulty buying. Stock moved rapidly, and I made several trips a day to the warehouse to keep the shelves and counters full.

But the really interesting part of my job was my helpers. Camp Shanks housed not only the returning soldiers but POWs from Italy and Germany captured, as I remember, in Sicily. Although the war in Europe was over, they were still held at Camp Shanks. I had two Italians and one German as helpers. The Italians were cheerful, fun guys who were enjoying being in the US. The German was quiet but never openly complained. All were awaiting transportation home.

Many of the Italians had relatives in New York City, and they would be taken there by bus or train for liberty. We would see them traveling down Tappan Road on weekends, laughing and waving out the windows of the buses. I kept the names and Naples' addresses of my two Italian helpers in my wallet for many years in hopes of one day meeting them again. Unfortunately, when I finally had the chance to visit Naples in 1951 as a young Navy ensign, their addresses were nowhere to be found. VJ Day was August 15, 1945, almost coinciding with my last days at Camp Shanks and the best job I had had until then.

Boating on the Hudson

On what, you may wonder, did I spend the money earned at Camp Shanks? I can't tell you how much I made, long forgotten, but it was probably less than a dollar per hour, perhaps $30 dollars a week. Part was spent on my next big project. A neighbor a few blocks away had a partly completed boat hull in his backyard. I had been eying it for some time, and with this infusion of money I convinced Harvey that we should buy it. We carefully inspected the hull and found no indication of rotting or damaged wood; it was definitely seaworthy in the imaginations of two 16-year-olds. I think my neighbor charged us $100 for his unfinished labor; it had been bottoms-up on blocks for a long time.

Our first problem was getting the hull to my backyard. It was 17 feet long, made of Philippine mahogany, and very heavy. We couldn't get the boat from the backyard to the front because of obstructions, so we cut a path through the woods from the backyard to Summit Street. Jimmy Frosco's father borrowed a two-wheel trailer used to carry long poles for the power company where he worked. We balanced the boat over the axle and slowly pushed and pulled it out to the street. Then Harvey's father hitched it to his car and pulled it the three blocks to my house. In my backyard we placed the boat hull-up on two heavy planks resting on cement blocks and contemplated how we would turn it into a beautiful boat capable of long voyages on the Hudson River.

The hull was sturdily built of planks four inches wide and a half inch thick, each running from the graceful bow to the narrow stern. The stern was a solid piece of thick wood to which we planned to attach an outboard motor. The planks were screwed to ribs and stems of unknown wood and appeared to be tightly attached. There wasn't any caulking between the planks. There was no keel, and the deck in the bow and running along the sides consisted of narrow wooden strips, also uncaulked. Our plan included finishing the hull, building a cabin and lower deck for sleeping, and then finding an outboard motor big enough to move the boat smartly along. Our ultimate destination was the Erie Canal! Who said 16-year-olds can't think big? What matter that neither of us had any boating experience? That's how we would get experience!

For the next two years we followed our plan, working when we could, learning and solving problems. From time to time my father made suggestions, but we did all the planning and work. I got a job the next summer at Lanigan's Hardware in Englewood that allowed me to buy at a discount the hundreds of brass screws of various sizes and other hardware needed to finish the boat. It takes a lot of brass screws to build a boat, and all the hardware, such as cleats, had to be corrosion-proof as the Hudson River contained brackish water. We bought a twelve-foot-long three-inch-thick pine plank that we shaped into a keel and attached with heavy bolts. Plywood for the cabin and deck had to be marine-grade so it could hold up against constant moisture. Caulking consisted of strips of felt-like material packed between the planks with a mallet and chisel.

We finished off the hull with days of hand sanding (no power tools in 1946). For the last touch we bought white marine paint and applied several coats. The bottom of the hull, from slightly above what we hoped would be the waterline to the keel, was painted green. Then we turned the hull deckside up, built a small cabin with two windows and a lower deck, and installed battery-operated red and green running lights.

We finished the boat in the summer of 1947, the summer before I started college. We scouted around and found a second-hand 9.7 horsepower Evinrude. I don't recall what we paid for it, but it must have been reasonable considering our overall financial resources. Based on our limited knowledge and advice from others, we believed it was powerful enough to move the boat at a good speed. We rented a mooring spot at a marina north of the Piermont Pier, made a mooring anchor from an old car engine, and chained a 50-gallon oil drum to the engine as a float. We loaded this into a rowboat and dumped it at our assigned spot.

In July we towed the boat to Piermont and launched. It didn't sink and took on only a little water that we assumed, correctly, would stop once all the seams expanded. We hadn't given it a name, but beforehand we had registered our "Cabin Cruiser" with the Coast Guard and the brass numbers (10U1392) were screwed to the bow. We towed it out to our mooring, cranked up the motor which we had run a few times for practice in a tank of water, and took a short cruise. Coming back we tied up to our float, covered the boat with a large tarpaulin, rowed back to the beach and admired our handiwork from a distance. We were now the proud owners of a beautiful luxury yacht with private quarters that came complete with sleeping bags and air mattresses.

Over the next days we practiced with a few short trips on the Tappan Zee and declared ourselves ready for the big adventure. We would sail up the Hudson to Albany, about 130 miles. Depending on time and finances, we might take a short trip on the Eire Canal. We thought we could make about five miles an hour motoring against the current. At that speed it would take three days to reach Albany including stopping for gas, unscheduled delays, and tying up somewhere for the night, then two days back sailing with the current. We estimated we could easily make the round trip, if all went well, in one week.

Off we went on a sunny morning, out into the mighty Hudson, sharing the river with ocean-going freighters. Were we scared? Only a little. We sailed north staying close to the eastern shore, dodging the big ships and their wakes that came at us from two directions after they bounced off the shore. We were making a little better time than expected; staying near shore reduced the current. We passed the Bear Mountain Bridge and West Point Military Academy a little after noon. We were deciding whether to tie up and have lunch or continue sailing while eating when our little 9.7 horsepower engine made the decision for us. It quit.

I took the carburetor apart to see if I could figure out why the motor had stopped. It had been running a little rough and, since the fuel was a mixture of oil and gasoline, I thought the carburetor jets might be clogged. I cleaned them up and put the top back on the carburetor. Bad news! In screwing it down I had cracked the top. I don't know about other Evinrude models, but our carburetor top was made of some sort of light-weight composite metal that we learned the hard way had little strength. When we restarted the motor with the cracked carburetor cover back in place fuel spewed out, a potentially dangerous situation if it ignited.

We were opposite the little town of Cold Spring, so Harvey jumped out with our mooring line attached and tried to pull the boat while swimming toward shore. After a few minutes it was clear that wouldn't work, so he got back in and we took turns using our one oar to paddle, with difficulty, to a dock on the waterfront. I went ashore looking for a shop that might have outboard motor parts but couldn't find one. What to do?

I had had a little experience working on my brother's Model A Ford motor, so I bought a tube of automotive gasket compound, intending to fill the crack and keep the fuel from leaking out. After applying this homemade remedy and allowing a short time for the compound to dry, we cranked the motor. It started up, but a lot of fuel still leaked out of the crack. Now we were in a real bind, over 25 miles from home with a motor we couldn't use.

We decided we had to return to Piermont, and the only way was to sail using the wind. We bought a twelve-foot 4x4 at a local lumber yard and stepped it in the boat aft of the cabin. Then we rigged our tarpaulin to the 4x4 and pushed out into the river. The current would help us but we might not get much of a boost from the wind considering that neither of us had any experience in using a sail. There was still a little daylight, but night was fast approaching and we took our chances that we could make a few miles that evening. The friendly current and some wind pushed us along, and we decided to head toward the west shore. We turned on our running lights and we were making pretty good progress. The Bear Mountain Bridge was just ahead.

While still north of the bridge we turned and looked astern and saw a large freighter bearing down on us. Clearly the watch on the bridge didn't see our tiny lights and never expected to meet a small boat on the river at night. Harvey grabbed our oar and frantically paddled toward shore. The ship missed us by just a few yards with the watch by this time finally seeing us and yelling to get off the river. Its wake almost caused us to capsize. Lesson learned, we decided we had to get off the river quickly and tie up.

We coasted along for another few miles, passing under the bridge and looking for a place to put in. We couldn't see any villages or private piers, only trees and scrubby underbrush. With no other alternative we paddled over to shore and tied up to a tree. We prepared to bed down and go to sleep, all the while talking about our close call. Suddenly a big searchlight beamed down on us. A soldier with rifle at the ready yelled at us and asked what we were doing. Wow! Talk about being scared! Where were we? We explained our predicament to the soldier; our motor was broken and we were just trying to get back home. We could tell he didn't believe us and said we had to leave. We described our close call with the freighter and asked to stay until daylight. He must have radioed to someone and agreed we could stay. However, for the rest of the night we had the searchlight focused on us and an armed guard standing by. We didn't get much sleep.

With first light we untied and pushed off with a guard still watching. I decided to try the motor again. I pulled the cord around the flywheel and it started right up. Apparently the gasket compound had hardened overnight. It made a better seal on the carburetor cover than the day before, and the fuel leak was small enough that we could collect it in a can and pour it back into the fuel tank. Looking back we could see the guard who was probably thinking we had made up our story about our problem. We returned to our Piermont mooring without further incident, tied up, and went home exhausted.

What was the mystery place where we spent the night under guard? With a little research it was easily determined. Murphy's Law worked to perfection. With all the millions of trees lining the banks of the Hudson River we had tied up to one at Iona Island, a US Navy ammunition storage depot with no trespassing allowed. Luckily we weren't shot.

Soon after leaving our beautiful boat tied up at Piermont I had to report to Columbia to start my freshman year, and we didn't get back for a few weeks. In early September a strong nor'easter blew in. We went to see how our boat had survived and found her beached and lying on her side. We couldn't right her and get her back in the water, so we left her on the beach planning to return in a few days. It was longer than that, and when we returned she was gone. The situation was very disturbing and strange.

When we first found our boat on the beach we saw that the rope hadn't parted, and I knew the type knot that we used could not have worked loose. Our anchor and float were still in their proper positions. No one at the marina claimed any knowledge of what had happened. We searched around but didn't see our boat in the water or in the front yard of nearby homes. I believed she was used when we weren't around and not tied up correctly. And, when we didn't return for a few weeks, someone claimed her as abandoned property even though the registration number was on the bow. Thus our small boat adventures ended abruptly with an unanswered question, but we both would soon move on to careers involving much larger vessels.

Applying for NROTC Scholarship

At the beginning of my senior year in high school I began thinking seriously of my options on going to college. It was clear I would need to apply for some type of scholarship, as I didn't expect my parents to provide the required financial aid. My grades were very good and, as mentioned, I had taken every math and science course offered as well as two foreign languages, the latter a requirement for consideration by Ivy League schools. I had participated in some after-school activities and committees and had success on the track team. I hoped all that background would be attractive to some colleges and perhaps earn at least a partial scholarship.

By chance my father read about a new Navy program, the Holloway Plan - Naval Reserve Officer Training Corps (NROTC), that paid for tuition, books and other fees at a college of one's choice. Graduates of the program would receive a commission as ensign. In return they would make a three-year commitment to serve on active duty. It sounded like a good deal, and I applied.

A short time after completing the required forms I was invited to a Navy office in New York City to take a written exam, probably the Navy's equivalent of the new SAT used by some colleges for admission.

Written and multiple choice exams were always easy and I must have done well, for a few weeks later I was invited to take a physical, my first but not last experience standing in line with a bunch of guys in just my shorts. I don't recall if there were any psychological tests or interviews. In January I received a letter saying I had been selected as an alternate to be awarded a scholarship. Alternate didn't sound too encouraging, but it was better than rejection. The next requirement was to apply to ten colleges that had NROTC units and send the responses to the Navy.

I applied to all the Ivy League schools plus Georgia Tech and Notre Dame and was accepted by all. I sent the college responses to the Navy and asked to go to Columbia. Then began another wait; in the meantime I finished my senior year at Tenafly and graduated in June, making the honor roll. (Of special note, our class had 22 veterans who had returned to earn their high school diplomas.) On July 15, 1947, I received the official notification that I had been accepted for admission to Columbia and was to report to the Professor of Naval Science "prior to 17 September 1947 to draw uniforms and to schedule your Naval Science Courses for the Academic Year."

Brother and Sister Update

My brother graduated from Tenafly in June 1943, before I started the following September. He is three years older but had skipped third grade and thus was four school years ahead of me. This four-year difference meant that his friends and daily activities were never the same as mine and we shared few interests during our teen years.

Tom had just turned 17 when he graduated from high school. The war was in full swing, so he would have to register for the draft when he was 18. In the meantime he took a job working as a draftsman with the New York Central Railroad in New York City. Learning of an Army program that would send him to college, he enlisted at the end of 1943 before he was 18. He was sent to Princeton for two semesters and then to Texas A&M. From there he went to Camp Blanding in Florida for basic training. As a private and after a series of short duty assignments in the US he arrived in Hawaii in July 1945. A month later, before he was sent to a combat unit, Japan surrendered. He would not be shipped home until April 1946, as Pacific theater combat veterans were given priority for the first available transportation back to the States.

Back home Tom enrolled in Cornell University's School of Hotel Management. On a summer internship in Grand Junction, Colorado, he met his future wife. After graduation they were married, and he landed a job as manager of a country club in Elizabeth, New Jersey.

The Korean War started in 1950. As he was in the Army reserve, he was called back to active duty. Now, with a college degree, he was commissioned a second lieutenant. With his background in hotel management he received orders to Frankfurt, Germany, where for the next two years he managed officers' clubs.

Discharged and back home again, he started work at a Savarins' restaurant in New York City. After a short time he moved his family to Grand Junction, his wife's home town, and became manager of the downtown La Court Hotel. This was at the height of the uranium mining boom in the West, and Grand Junction became a hub for those involved. Tom tells many lively tales about those times, including owning his own mine near Moab, Utah. This brief history is provided in case you were wondering what happened to him. He will have to write his own autobiography if you want to learn more about his three daughters and two sons from his first marriage.

Before our teen years my sister and I had been quite close. Marge answered to the name Babe for many years before deciding it wasn't dignified. We did many things together as there were few children in the immediate neighborhood. We built tree houses, and we were free to explore the surrounding area that was largely empty lots and woods. But by the time I entered high school she had her circle of friends and I had mine.

Marge graduated from Tenafly three years after me and attended Syracuse University. When my parents moved to Fair Haven, New Jersey, in 1951 she met a local young man, Richard "Turk" Berger. While I was still in the Navy they were married, moved to Squantum, Massachusetts, and began to raise a family of four girls. Marge spent many years researching our family history, traveling to Ireland, visiting cemeteries and collecting information from relatives, but the notes she recorded could not be found after her death in 2007. A sad loss; perhaps one day it will be found. Turk and their daughters will have to fill in the blanks of her life that I have not included in this story.

First home on Staten Island, 1930

Grandmother Muz, mother holding me with Tom in backyard, 1930 (left)
Tom and I in backyard of first home, 1932 (right)

Second home on Staten Island, 1932, father with Tom in wheel barrow (left)
Mother holding my sister Margaret, Tom and I in back yard, 1934 (right)

Another view of home

Father in uniform, 1918, WW I Victory medal with Escort bar

USS Beaufort - AK-6

Father on liberty in Chicago, 1918

Mother's family. Back row, James, Helen, Betty and Mark. Front row, Lucille, Muz, Pa and Loretta (1940?)

Norwood home

Mother, Tom and I on front porch, sister hidden behind my mother, 1937

Father with Lassie, front porch, with Tom and me, sister just behind my father, 1937

Norwood's public school in 1940

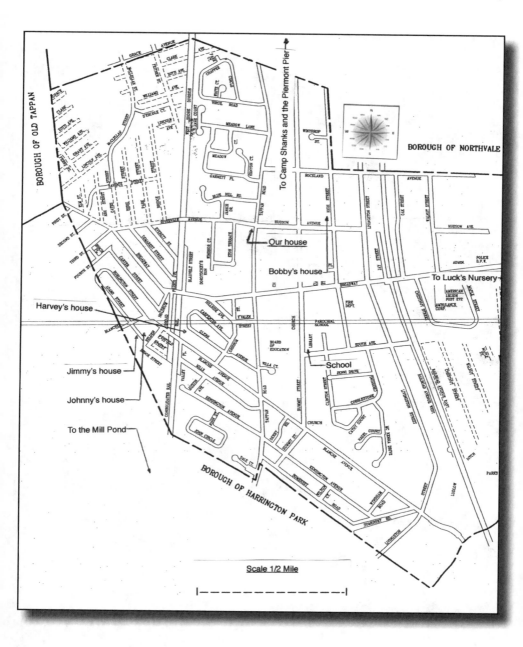

Map of the west side of Norwood. Some streets shown didn't exist in the '30s

Camp at Black Rock State Park, Bill Lucien on left, 1939

1943 graduating class picture sans me. Marie Luck center girl middle row.
Back row, second Barry Sweeney, fourth Harvey Terwilliger, last Jimmy Frosco.
Middle row, second Richie Vogler, second from end Bobby Sticco

1932 Dodge - not the original

1937 Oldsmobile - not the original

High school yearbook photos, Mary Beech and me

Bobby Sticco and Harvey Terwilliger

Chapter II
Columbia College 1947 to 1951

On September 24, along with 45 other freshmen, I was sworn into the U.S. Naval Reserve as a midshipman. So began my Navy career which, with several interruptions, ended in 1967 when I retired from the Navy Reserve with the rank of lieutenant commander.

When I reported to the NROTC armory at Columbia in September, I was issued a new dress blue uniform, navy-blue flannel shirt, dress-white pants, an officer's hat with a fouled anchor emblem and white and navy-blue covers, gray gloves, a navy-blue tie and a hand-me-down black overcoat. All were worn for the next four years, getting a little snug as I grew. I had to buy the required plain-toed black shoes from the $50 monthly stipend I received. I already had the required white shirt. I thought I looked so good in my uniform that the Barnard girls across Broadway from the Columbia campus would all go crazy. However, they must have had other important things to do and stayed in their dorms because I never met any crazy Barnard girls in the next four years.

Freshman Year

My first year at Columbia was, in many ways, very different from what would be my last three. Because my financial condition was not exactly robust, I decided to live at home. The monthly $50 from the Navy paid for incidental expenses. Commuting from Norwood by bus and subway cost a little over a dollar a day, which left money for lunch at the school cafeteria and a few other "luxuries." Many students commuted, living as I did at home somewhere in the New York metropolitan area. The Seventh Avenue subway stopped at 116th Street, at the entrance to the campus. This wasn't a bad way to attend college because I used the one hour it took each way to do homework. On Fridays I felt a little self-conscious at first because that was drill day and I had to wear my uniform. The war had ended two years earlier, and one seldom saw a person in uniform. However, I never quite got over the feeling that everyone was looking at me.

With boundless energy, in spite of the long daily commute, I began to participate in some extracurricular activities. The first was a part of freshman hazing. Freshmen were required to wear a small Columbia-blue cap for the first months. One of the hazing rituals was placing a cap on top of a tall, thick pole planted in South Field. If the freshman class could figure out a way to climb the pole with no mechanical help and remove the cap, the requirement to wear the silly looking thing was rescinded. The pole must have been at least 20 feet high and was covered with a heavy grease. No freshman class in Columbia's 193-year history had ever succeeded in removing the cap.

The class of 1951 became the first to accomplish the impossible. A quickly assembled gang with several large NROTC freshmen (Ollie Van den Berg, Jay Dee Battenberg and "Doc" Sylvester, all on the freshman football team) among those at the bottom locked their arms around each other and the pole and hung on. I was in the next tier, standing on Doc's shoulders, also holding on to the pole. Others stood on our shoulders, and we built a pyramid of yelling guys. Little Al DeBertolo climbed over all of us and grabbed the cap. As far as I know no other class has accomplished this amazing feat. We should all have been immediately awarded degrees in social engineering without attending any further classes! I wonder if freshmen still have to try to climb a greased pole. Probably not. It wouldn't be politically correct, as Columbia College now enrolls women who probably wouldn't want to participate in such a messy event.

Naval Science courses during the four years, covering seamanship, naval history, navigation and many other subjects, were taught in the Armory a few blocks west of the main campus. We also drilled in the Armory on Fridays if the weather didn't permit using South Field on the main campus. South Field, a few acres in size, had felt the tread of many marching feet through the years. Columbia was the alma mater of thousands of Navy midshipmen turned ensign during WW II. Because of its high usage it was oiled to keep the dust down, and when wet it was a gooey mess, especially hard on shiny black shoes. In the winter a 220-yard oval wooden track was installed for track team practice that reduced the size of the drill field.

Intramural meets were also held on this track. In our freshmen year several of us in the NROTC entered. Merritt Rhoad and I won our events, the mile for Merritt and 440 for me. This led track coach Carl Merner to invite us to join the scholarship holders on the track team, which we did, running at first with the cross-country team. A year later Merner offered me a track scholarship. As I already held a scholarship which would allow me to pursue an exciting Navy career, assuming I would get through the next three years, I thanked him and declined.

Along with the freshman courses required for a Columbia degree and for Naval Science, I took courses in geology and archaeology. I had no desire to be chained to a desk in order to make a living, so after serving in the Navy a possible career in either of these fields was appealing.

In our high school yearbook I listed as my ambition to be an archaeologist. Archaeology was offered only in graduate school. I wasn't bashful; I made an appointment with the department head, Dr. Dinsmore, and convinced him of my interest and ability to take some of the courses offered. For the next two years I sat next to much older students who were majoring in archaeology. I more than held my own, receiving straight A's for my work. I still have one of my A papers on the use of aerial photography to locate ancient sites. Through the years my interest in matters archaeological has remained strong. I have visited many important sites around the world, some with family as you will read.

My first geology course was taught by renowned geomorphologist Armin K. Lobeck. It was an introductory freshman course for those intending to go on in geology, and it could fulfill a "science course requirement" for non-science majors. It was fun and easy, and I always looked forward to his lectures. He was a portly, distinguished looking gentleman sporting a gray goatee. He started each class standing in front of blank blackboards that stretched across the front of a large auditorium. By the end of the lecture the blackboards were filled with exquisite chalk drawings covering the day's subject. I wish I had had the technology now available to all students, a cell phone camera, to capture his drawings. They would have made a remarkable record of Lobeck's lectures.

Geomorphology course work also included several local field trips that Lobeck led with the assistance of graduate students. We went to the Palisades in New Jersey, Long Island's glacial terminal moraine, and other locations demonstrating some geomorphological feature he had described in earlier class lectures. His knowledge and ability to explain complicated geological processes provided the inspiration that led me to more geology courses at Columbia and eventually to an advanced degree in geology ten years later.

As part of the core curriculum all undergraduates were required to take eight semesters, spread over three years, of two courses structured to familiarize us with authors who influenced Western culture titled Contemporary Civilization and Humanities. They were listed in the course catalog as providing students "with the opportunity to develop in small seminar settings intellectual relationships with faculty early in their college career." With these courses under our belts Columbia believed it would send into the world well-rounded graduates who would eventually become leaders in industry, academe, and government.

I found the thrice weekly classes and preparation for them very time-consuming. They were led by professors, some of great renown, who analyzed the meanings of each author's works in excruciating detail and attempted to show how relevant they were to the present day. The "small seminars" usually resulted in a few students trying to attain an "intellectual relationship" with the prof by dominating the discussion while many of us listened in bored resignation. There's no question philosophers and writers we studied, such as Sophocles, Plato, and of more recent vintage Spinoza and John Locke, greatly influenced Western thought and our nation's founders. But by this time my conservative genes were in control and I objected to analyzing Marx and weird German philosophers such as Hegel and Nietzsche. And who needs to waste time studying Rabelais? Some classmates thought the courses were great, but for three years I suffered through the lectures and tests, receiving mediocre grades that reflected my general lack of enthusiasm.

All things considered, at the end of my freshman year I thought it had been quite successful. I had reasonably good grades, had been active in athletics and NROTC-sponsored events, and had made many new friends. During Christmas and Easter breaks, to make a few bucks, I continued to work at Lanigan's new hardware store in Dumont.

I saw my friend Harvey from time to time while at home and told him about all my activities. He did not have any college plans at the end of high school and also needed a scholarship to continue his education. I learned that Kings Point Merchant Marine Academy was offering scholarships and suggested he apply. At the end of four years graduates would receive a commission as ensign in the Navy or they could go to sea on a commercial ship if they passed the exam for third mate. Harvey applied and was accepted. After graduation he went on to a distinguished career, serving first in the Navy and then in the Merchant Marine, finally becoming captain of several large container ships. On one of his voyages, while unloading and picking up cargo in New Zealand, he met his future wife Pat.

First Midshipman Cruise - June 1948 to August 1948

The culmination of freshman year was my first NROTC summer cruise that began the week after last classes in June, a key event for freshman midshipmen to test our knowledge and suitability to continue to a commission. We had all been looking forward to it with great anticipation. The dress uniform, except for the black shoes, was left behind. In its place we were issued two sets of enlisted men's summer-white uniforms, black neckerchief, two hats, two sets of blue dungarees, underwear, white socks, and a duffel bag. The only difference between our uniforms and the enlisted crew was our hats. Ours had blue rims to distinguish us from the ship's crew on our assigned ship. There would be no way to hide by mingling with the crew. Before leaving we were given stencils of our last names which we marked in black ink on each piece of clothing.

With some of the senior midshipmen we boarded a bus from Columbia to Pennsylvania Station. By train we went to the naval base at Norfolk, Virginia, a long day's journey. We were housed in enlisted men's barracks for two days before going onboard our ships, the heavy cruisers USS Macon CA 132 and USS Columbus CA 74. The Macon, commissioned in 1945, was one of the Navy's most modern ships. The Columbus was a few years older.

NROTC midshipmen from Columbia and Harvard were selected to join Naval Academy plebes, also on their first cruise, on the Macon. We had been divided into two groups, alphabetically, and about half of us with the first letters of the alphabet in our last names assigned to the Macon and the other half to the Columbus. First-year NROTC midshipmen from all the other schools around the country were assigned to the Columbus. Senior Annapolis midshipmen performed as junior officers, sometimes telling us what to do. I didn't keep a journal during the cruise, but dates and events that follow are accurate to the best of my memory.

Our first full day in Norfolk consisted of tours of the base and a stop at the base ship's store to buy last-minute supplies. In the afternoon, assembled on the dock, we were treated to an air show by the Blue Angels, led by their original organizer LCDR "Butch" Voris. The team was now flying F8F-Bearcats, having switched the year before from F6F-Wildcats of WW II fame. The Bearcat was a real bear. In 1946 it established a world record that wasn't broken for many years, climbing from takeoff to 10,000 feet in 94 seconds. It was a small plane powered by a Pratt & Whitney R-2800, 2000 horse power, radial engine. Years later, pilots flying surplus Bearcats in air shows would take off from the runway, immediately do a loop, and land back on the runway -wow! The Blue Angels put on an amazing show of close formation flying. It was the first time I had seen them perform.

One of the reasons for the show, I assume, was to impress Annapolis second-year midshipmen who were starting their aviation training summer cruise on the USS Coral Sea, CVB-43, berthed nearby. Watching the Blue Angels probably reinforced the desire of some midshipmen to apply for flight school; it certainly started me down that path.

On June 7 we cast off all lines and I began my first sea cruise, manning the rail with my new shipmates, standing at attention in our new white uniforms. The band was on the fo'c'sle playing Anchors Aweigh, and families of the crew were on the dock waving goodbye. It was a very stirring event accompanied by some trepidation on my part. Would I get seasick? Would I adjust easily to life at sea? Annapolis midshipmen were accustomed to the daily grind of tight schedules and discipline. We NROTC acolytes had led much freer lives. Our destination was the Mediterranean Sea to rendezvous with the US Sixth Fleet.

On board our quarters were the same as those of the enlisted crew, three-high folding bunks suspended on chains with thin mattresses and a small locker to stow our gear. Thirty or 40 midshipmen were assigned to each berthing quarter, NROTC and Annapolis midshipmen mixed together based on alphabetical last names. A period of adjustment was needed to meld our different cultures, but as time went by we became good shipmates. We all strove to adapt and excel, although some from Annapolis considered NROTC midshipmen to be second class citizens. Fitness reports would be written and filed after the cruise showing whether or not we had succeeded.

We ate in the mess hall with the enlisted crew. The food was passable. My one lasting memory is of the strong sulfurous smell of the eggs being cooked at breakfast. We were told the Navy was using up millions of eggs that had been stored in caves during the war. This sounded reasonable, but after that I didn't eat eggs again for a very long time. When it was open for a few hours each day we could go to the ship's "gee dunk" stand to buy candy, ice cream, or some other treat. I don't know where the name came from, but it must derive from the old Navy.

As the cruise progressed we rotated through many different duties from the bridge to the engine room, from forward anchor chain stowage to steering aft. We stood standard Navy watches, four hours on and eight hours off. I don't recall my first duty station but clearly remember an early initiation into two Navy traditions, holy stoning the deck and chipping paint.

Holy stoning is a throwback to the days of wooden ships and iron men. On the Macon it consisted of crawling backward on hands and knees, sans shoes and socks, as a bosun's mate splashed sea water on the teakwood main deck and we rubbed it with large pumice stones. The Macon's main deck stretched 674' 11" from the bow to the stern. When wet the teak had a light brown color, but when it dried it was a brilliant white. It was the welcome mat for visiting dignitaries, demonstrating the Navy's penchant for cleanliness. There we were, rows of midshipmen, side by side crawling down the deck while a bosun's mate enjoyed himself making uncomplimentary remarks about our holy stoning skills, a great opportunity to torment future officers who would soon require a salute.

Ships built after WW II no longer have wooden decks. It was found that combat damage resulted in added casualties from deck splinters and it complicated making repairs. Midshipmen are now spared the ignominy of crawling around while a bosun's mate cracks the whip. However, for our generation it was a useful lesson in humility although hard on the knees.

Chipping paint was the other manual labor inflicted on future officers to demonstrate our lowly position as midshipmen. This was not a throwback to the days of wooden ships but a more modern labor after wood gave way to steel. Sea air and salt water continually conspire to turn ships of steel into big lumps of rust. Keeping a ship at sea in fighting condition and looking good for port calls meant constant chipping and cleaning of rust spots, followed by painting. We would examine the area assigned and with a steel mallet chip a paint blister or rust spot down to a shiny surface, then paint the cleaned area with a yellow primer. When dry, the primer would be painted over with battleship gray. This labor surely was the origin of the saying: "If it don't move, paint it." On paint chipping days the ship sounded as if a thousand woodpeckers were attacking it, making a boson mate's heart fill with joy. If chipping took place while we were trying to sleep after coming off watch, good luck. Although we participated in paint chipping, it was also a form of punishment for crew who went before a Captain's Mast (judge and jury) when caught breaking the rules. Keeping the ship looking good was a constant struggle.

As we steamed across the Atlantic heading for our first liberty port of Lisbon, Portugal, we all looked forward to learning to operate and fire the ship's guns.. The Macon had a main battery of nine eight-inch guns in three-three barrel turrets, two forward and one aft, six dual mount five-inch gun turrets located along the sides, 46 40 mm antiaircraft guns in 23 dual gun mounts in every corner of the ship and 28 20 mm mounts. Most of our work on the smaller guns was as munition passers and loaders. However, I was selected as a pointer in an eight-inch gun turret. Another midshipman sat on the other side of the turret as the trainer and turned the turret to the required position. I would match pointers on the gun control display in front of me to elevate the gun to the correct angle, then on command I pulled the trigger to fire the guns.

Shells and powder bags were hoisted into the turrets on conveyors from magazines far below deck. Loading was done by a hydraulic ram while the gun barrel was horizontal. Powder bags were pushed in behind the shell and the breech closed, each step initiated by a gunner's mate. Then I would do my job of matching pointers, raising the barrel. The trigger was on my hand controller. We were to fire on a towed target sled thousands of yards away. On the first firing, as I was matching pointers, the gun discharged prematurely, causing great consternation. Fortunately the turret was pointed in the correct direction The drill was stopped, and we never fired the eight-inch again. We were never told the reason for the premature firing.

As standard equipment for heavy cruisers, the Macon carried two seaplanes that could be launched from catapults on the stern. They were WW II vintage aircraft carrying a pilot and one crewman. Their mission was to act as scouts flying ahead of the task force. As we sailed east toward Europe we were given a demonstration of a launch and recovery. The recovery was the most interesting maneuver. As the plane returned to land, the ship began a slow turn of 180 degrees. This created a wake inside the turn that reduced wave height. After landing in the calmer water, the plane taxied onto a large cargo net towed by the ship. It was winched alongside and then lifted on board by the crane on the fantail. In the modern Navy seaplanes have been replaced on cruisers, frigates and smaller vessels by helicopters. This eliminates the time-consuming, complex recovery maneuvers that would slow the task force and leave it vulnerable to submarine attack. Over the horizon scouting is done by aircraft flying off carriers.

As mentioned earlier, one of our rotated duty stations was in steering aft. This was a small enclosure at the stern far below deck and connected by sound-powered phone to the bridge. Most of the phone system throughout the ship was sound-powered since battle damage could put electric generators out of commission. Gun turrets, the engine room, and other key facilities were connected in this way. Steering aft had a backup helm to assure that the ship could be steered if damage to the bridge resulted in not being able to steer the ship from there. To make sure the duty crew was on station and all systems operating, periodic voice checks were made. Occasionally, during an emergency drill, the ship would be steered for a short time from this station.

Near steering aft, but farther below deck, was another duty station, shaft alley. Below it were only the ballast and fuel tanks. In shaft alley the ship's propeller shafts connected the two propellors outside the hull to turbines in the engine room. It was a hot, sweaty, noisy place reeking of the smell of oil from the huge, rapidly rotating shafts. Sailors were stationed here to monitor temperature gauges on the large bearings holding the shafts, including the final bearings where the shafts penetrated the ship's hull to assure that they were continually lubricated and not overheating, a most critical task.

As we steamed east we learned from our Annapolis shipmates a new ritual, spit-shining our shoes. To practice this important rite of passage we would sit on our bunks, small rag in hand, and apply coat after coat of black polish lubricated with a little spit. The result, once we mastered the technique, was a brilliant shine that made an inspecting officer beam with pleasure. After a week at sea we arrived off the coast of Portugal. As we entered Lisbon harbor we manned the rail in our dress white uniforms and shiny-toed shoes. Unfortunately the spectators on shore couldn't appreciate our new expertise.

Cruisers and destroyers steamed up the Tagus River and tied up not very far from downtown Lisbon. The Coral Sea, because of her deeper draft, was forced to anchor offshore. We were all eager to take in the sights. Our meager finances didn't allow most of us to sample the casino and night life, but a low-cost trip to Fatima was offered and we went by bus. It felt good to get my land legs back as I walked around the city, ate some local food, and bought a few trinkets to take back home. During the four days tied up to a dock we had only a few hours of liberty, and then we set sail again.

Our training routines continued as we passed Gibraltar and entered the Mediterranean. Our next port of call would be Livorno (Leghorn) on the northwest coast of Italy, home of the Italian Naval Academy. A few days after leaving Lisbon we were called up on deck for an unusual "short-arm" inspection. Apparently some midshipmen had reported to sickbay with an exotic disease common to sailors around the world. We opened the fly on our uniform pants and pulled out our male member. When the medical corpsman or doctor walked by we had to show whether or not we had picked up one of the diseases, today identified by the acronym STD. Some shipmates with a little extra cash had purchased unwanted gifts. How many had the problem I don't know; I wasn't one of them.

Steaming into the harbor at Leghorn at slow speed was a new experience. While manning the rail we were greeted with a multi-gun salute. We must have been an impressive sight to those on shore, a large flotilla of mighty ships representing the most powerful nation on earth, a former enemy turned friend. We had been briefed on proper protocol and representing our country with dignity, a sober message for a group of 18-year-olds.

Once again we were able to tie up near the city and walk downtown, or what was left of it, in a few minutes. Leghorn, an important port, had suffered major destruction during the war. Many buildings were still in ruins awaiting reconstruction. However, the streets were all cleared and the citizens seemed friendly. They welcomed us in anticipation, I guessed, of the money we would spend.

We didn't have a lot of money to spend, but we quickly learned that American cigarettes were in great demand by the locals. At the ship's store a carton of cigarettes cost about $2:00. This allowed us to participate in the great game of bartering. If a trinket seller had something we wanted, how many cigarettes did he want in exchange? We exchanged cigarettes for meals, and a few of us even went to a local vaudeville theater. We sat in the back, but we were easily observed as we were in uniform. Comedians on the stage apparently made a few jokes at our expense. People in the audience would turn, look at us, and laugh. We didn't understand Italian but knew we were being made fun of, so we laughed too. A day of liberty cost only one carton of cigarettes! While in Leghorn we took a short sightseeing trip to Pisa and Florence and pictures I took of the leaning tower and statues in the Uffizi gallery survive.

We left Livorno on a beautiful sunny early July morning, manning the rail until clear of the harbor. The Mediterranean was warm, blue and calm as we headed west for our next destination, Gibraltar. My new duty station was one of the powder magazines far below deck. It was not a pleasant place to stand watch as the powder bags, stowed neatly all around the walls, gave off an unpleasant odor. A few nights later we encountered a fierce storm, called a mistral, that blew off the French coast and sent the ship violently heaving up and down and rolling from side to side.

The outcome was preordained; we landlubbers all became seasick. There was no escape; we had to finish our watch suffering the smirks of the ship's crew on watch with us. Fortunately there were buckets available and soon, in addition to the sickly powder smell, the stench of vomit filled the stuffy air. It was a terrible four-hour watch, the worst I ever endured in my Navy career. It left me weak and barely able to make the long climb up ladders to finally breathe fresh air, my first but not last exposure to what a rough sea can do to one's equilibrium although later, on active duty, I became adjusted to rough seas. For those who have never experienced seasickness the old saying goes, "You think you are going to die, and after awhile you hope you will."

Our reason for going to Gibraltar was to change ships. Annapolis midshipmen were scheduled to stay at sea a month longer than the NROTC cruise. Annapolis midshipmen on the Columbus would transfer to the Macon. We NROTC midshipmen on the Macon would go onboard the Columbus which became a training ship for NROTC only. Annapolis midshipmen would sail to the Caribbean while we would take a more direct route back to Norfolk.

We were in Gibraltar only a few days while the transfer took place. On shore we toured the "Rock," including some of the caves that held large-caliber guns. We even saw a few Gibraltar Barbary apes climbing around. We had tied up to a very long pier, the Macon being last in line. To go ashore we walked up the pier past several British ships, also tied up. Although the British Navy has a proud history, it seemed that careful personal cleanliness was not part of the tradition. When we got alongside the ships we could smell uniforms and bodies not washed very often. I guess British sailors became accustomed to the environment and didn't notice it. It probably saved on soap and a ship at sea's precious commodity, fresh water. If this book is read in the future by British sailors, I hope it will not destroy our two countries' special relationship.

And so we left Gibraltar on the last leg of our cruise. A week more at sea and we were back at the Norfolk Naval Base, fully trained and seasoned salts. It had been a terrific experience for one who had never been anywhere on the water except in a small boat on the Hudson River, a ferry or the Circle Line. The incoming freshman class of NROTC midshipmen would have to be impressed by our worldly experience, tales, and great talent to spit-shine our shoes.

Sophomore Year

At the end of my freshman year I had come to the conclusion that commuting was too time-consuming and decided to live on campus. For example, to attend practice at Baker Field as a member of the track team I took an afternoon bus provided by the school that waited on the campus at 116th Street. Baker Field was at 218th Street in northern Manhattan. After practice I could take the same bus back to the campus if I lived there. But because I had to make a connection with a commercial bus back to Norwood, I took my books and homework assignments with me and after practice took the subway from the Baker Field station to the 168th Street bus terminal. Sometimes this arrangement became a little complicated and often, if practice ended late, I missed the frequent rush-hour buses and wasted time waiting for the next one. I arrived home in early evening, tired and hungry. To ease the financial burden of living on campus I had requested a job in the cafeteria and was accepted. My compensation would be a meal ticket reflecting the hours worked. With this added income I figured I could afford to live on campus.

I joined two other NROTC classmates, Merritt Rhoad and Herman Brown, who had roomed together during their freshman year. We moved into Hartley Hall, a dormitory on the east side of the campus, and were fortunate to get a third floor, two-room corner suite and split the rent three ways; single rooms held only one or two students. The main room was furnished with a double-decker bunk, a single bed, armoire, desk and chair. The other smaller room had two desks and chairs, a closet and wash basin. I won the toss for the single bunk. We also had a window facing the campus Quad. Showers and toilets were down the hall. On the ground floor were a couple of washers and dryers and a TV room.

Columbia College wasn't coed then, so no girls were allowed. Even visits were not permitted. We never thought this a problem; I don't believe any of the Ivy League colleges were coed and classes were all boys. Barnard College, across Broadway, also wasn't coed, only girls. If we felt the need for feminine companionship it was just a block away. Anyway, my roommates and I didn't have much time for girls; although some guys undoubtedly found the time.

Hartley Hall was adjacent to Hamilton Hall, where many of my classes were held. As an early riser I selected courses that began at eight o'clock, leaving afternoons free for activities such as track practice. I could roll out of bed and be at class in a few minutes. My roommates were not so predisposed, so I could dress in the other room and not have to share the small wash basin. Another advantage of living at Hartley Hall was that the campus was honeycombed with tunnels that connected utility services to most of the buildings. We could enter a tunnel in the Hamilton basement and walk to classes in other buildings in shirt sleeves, even in winter. Most students weren't aware of the tunnels or didn't use them, but I thought they were a great convenience.

At the cafeteria it was easy to accumulate as many hours and meal tickets as I was willing to work. I punched in and out on a time clock and the hours grew. I sold excess meal tickets to friends. Depending on my schedule I worked at everything from running the dishwasher to serving at the football training table.

I developed a good working relationship with the woman in charge of food services, and from time to time she asked me to serve at dinners she catered at the Columbia University Medical Center. This was a good deal as pay was a little better. Travel time counted from the campus to the hospital and back. I would put on a clean white shirt and take the subway to 168th Street. Dinners were held in a small dining room and usually included 20 or fewer guests. Then I went back to Columbia on the subway to late study hours in my room. Eventually, on Sunday evenings in my senior year, I managed the Lion's Den, the on-campus equivalent of a soda fountain that served the usual fare of milkshakes, hamburgers and other fast food with a jukebox, a few tables and chairs and a small dance floor.

Track became my major extracurricular activity. The varsity team was small, so I was able to compete in the 440, 880, mile relay, and occasionally the 1000 (yards, that is). We ran in meets competing with teams from Ivy League and other eastern colleges. I stopped competing in cross country after my freshman year; I was a sprinter, not a long distance man! Besides, cross country workouts were long and grueling as we trained by running up and down hills in Van Cortlandt Park.

Our only indoor track facility was in University Hall. It consisted of a balcony that circled the gym below and was used in bad weather. Otherwise in winter we used the wooden track that was set up in South Field. In the spring outdoor practice was held on the track around the football field at Baker Field, a standard 440-yard track with a 100-yard straightaway. College athletics in the 1940s were much different from those in the next decades. Funds to support teams were usually in short supply, and the whole approach in the Ivy League was to support true student athletes. Few went on to professional teams, Lou Gehrig and Sid Luckman being two notable Columbia exceptions. I imagine that revenue from tickets to the different team games never was sufficient to pay for equipment, travel and other expenses; and not even football turned a profit. There were no TV and equipment maker endorsements to help make ends meet.

We had a rudimentary weight room that I never recall using; maybe that's why my times didn't improve. But I don't remember coach Merner ever telling us to lift weights. The field house at Baker Field was old and drafty with a few offices, lockers, showers and dressing room. A more modern field house was built in my senior year. Columbia supplied warm-up suits, uniforms and two types of track shoes, one with short spikes for indoor races and the other with much longer spikes for outdoor cinder tracks. We never used starting blocks outdoors, and I don't remember using them in the 440 indoors. We used a small trowel to dig holes in the cinder track for a better start. Indoors we used a standing or kneeling start. If we fell on the wood track we received a few skin burns, and the trainer spent time taking splinters out of our extremities. We had a trainer, a great little Dane, who massaged our legs before and after running. Crew used a small tank below Low Library in the winter. My roommate Herman was on the lightweight crew. In bad weather during spring practice the baseball team had an indoor net batting cage in University Hall. That was the extent of our facilities that I remember, but I expect most Ivy League schools didn't offer much more.

Although I dated a few girls my freshman year, one at Vassar, I didn't have a steady girlfriend. I went out a few times with Bobby Sticco's stepsister Isabel who was a professional dancer and entertainer on cruise ships. On weekends we went to Bear Mountain Inn or local New Jersey night clubs that featured big bands and she patiently taught me the difference between my right and left feet.

However, during a weekend spent at Merritt's home in Philly, his girlfriend arranged a blind date with one of her classmates, a senior at Overbrook High School. She was a very pretty brunette who lived with her divorced mother in a large home in Chestnut Hill. It was a very compatible match, and whenever I could get to Philly we had a date. I even got along well with her mother, so the romance flourished and I would stay at their home over the weekend. But my sophomore year was filled with so many activities that our dates were infrequent.

By now I was a full member of the track team. Meets were held, usually on Fridays or weekends, at Columbia, other schools, Madison Square Garden, and armories in the immediate metropolitan area. The team, perhaps 20 guys, included a few very good athletes. Bob Solberg, our high jumper and discus thrower, regularly cleared 6' 6", a very good college height in 1949. Bill Holland was our 440-yard dash champ who could clock 48 seconds. Our 100 and 220-yard dash men were also very competitive. Doc Sylvester was our shot-putter. I ran the 440 and 880. But, despite my talented teammates, we were not very deep and we didn't do well in dual meets.

Despite this shortcoming, competing at the Penn Relays in the spring we had a memorable result. Running the second leg on the mile relay, I passed off to my teammate in good position for the third leg, who passed off to Holland, our anchor man in the middle of the pack, for the last leg. Bill ran a great last lap of 47.6 seconds, and we came in third with a time of three minutes and 21 seconds. I earned a bronze medal that I gave to my Philly girlfriend on our next date as a token of my affection. I wonder if she still has it.

Sophomore year ended without problems. Courses were similar to those I had taken as a freshman. I added Spanish since a foreign language was required to receive a degree. I took two Spanish courses, including one in my junior year, to fulfill the requirement. Grades were so-so. Counting both semesters I had seven C's (two C-plus), four B's (two B-plus) and two A's, one in my graduate anthropology course. Graduate students were "easy" to compete against. Four of the C's were in Contemporary Civilization and Humanities, reflecting my disinterest. In only one more semester I would be finished with those courses. In May I was awarded a varsity letter for my participation on the track team. I received a white cardigan with a Columbia-blue "C" on the front. It now hangs in the closet with other out-of-date and "too small to wear" garments reminding me of better days. The major event came again at the end of the school year, our second summer midshipman cruise.

Second Midshipman Cruise - June 1949 to August 1949

Our second cruise was designed to expose us to two new aspects of potential Navy careers. The first six weeks were spent at Pensacola, Florida, in a short introduction to naval aviation, followed by six weeks at Little Creek, Virginia, participating in amphibious operations. In addition to our dungarees, this year we were issued wash-khakis to wear in semitropical Florida and later in Virginia. They came with an overseas cap colloquially called a piss-cutter, complete with a fouled anchor insignia, very jaunty. We boarded a crowded train again at Pennsylvania Station and spent the next two days en route to Florida.

"You leave the Pennsylvania Station 'bout a quarter to four.
Read a magazine and then you're in Baltimore.
Dinner in the diner, nothing could be finer
than to have your ham and eggs in Carolina."

Our trip wasn't as cheery as the song; we weren't going to Chattanooga, and I can't recall how we were fed. We read a lot, did some "group singing" from a limited repertoire, and pulled a few practical jokes on unsuspecting dozing classmates. It was a long two days, passing slowly through cotton fields, poor and dilapidated parts of the South, and small towns. Most railroad tracks in the South don't go through the scenic parts. We slept the one night sitting in coach seats, but when you're 19 it really doesn't matter. Our destination was sure to be much more exciting. We were actually scheduled to fly!

Upon arrival we were joined by NROTC units from all over the country. This time each school's unit was housed separately in wooden barracks. We marched as a school unit to all classes and activities, trying to look very military and to stay in step. At Columbia we drilled only once a week and, depending on the weather, might not march for several weeks at a time and then for only about one hour. We selected Jim Evans, an enlisted man before becoming a midshipman, to lead our detachment. He would sing out a cadence as we marched along trying to look just like Annapolis midshipmen. There was informal competition between school units as to who marched the best, and when we marched near the unit from Rice University we would sing in cadence: *"What comes out of a Chinaman's ass? Rice! Rice!"* I don't recall how they retaliated, but I'm sure they had an equally vulgar comeback. If we had similar chants for other schools, they are now forgotten. We studied aircraft engines, structures, communication, navigation and more. Athletic events and competitions were also scheduled as team sports. We spent free time at the beach and the base movie, but I don't recall any memorable liberty in town.

Pensacola, founded by the Spanish in 1559, had witnessed many changes in its long history. Abandoned soon after the first settlers arrived, it was occupied off and on by the Spanish, then the French, then the British, and was finally purchased from Spain in 1819. The Civil War also had an influence. The downtown business area was not very large, and some businesses did not welcome Navy personnel. Nearby residential neighborhoods included many large homes dating from its early history. In 1949 Pensacola had a small-town atmosphere with one main industry, the Navy.

A short history of Naval aviation may be useful. After the Wright brothers demonstrated their Wright Flyer before a military audience at Fort Myer, Virginia, in 1908, both the Army and Navy became interested in the aeroplane's application for military use. In 1909 the US Army Signal Corps purchased a Wright flying machine. In 1911 Eugene Ely, a Curtiss pilot flying a Curtiss biplane, demonstrated a takeoff from a wooden platform built on the bow of the cruiser Birmingham. Later that year he landed on a temporary deck built on the stern of the battleship USS Pennsylvania anchored in San Diego harbor. Although Ely's demonstrations showed the potential to take off and land on a ship, the Navy at the time decided this wasn't practical and that seaplanes were a better fit for its operations. However, it was important to keep up with the Army; service rivalries could not be ignored, and since it operated on the oceans the Navy's first purchase was a Glenn Curtiss seaplane, the A-1 Triad. It was built at the Curtiss plant in Hammondsport with the purchase naval aviation began a history of extraordinary achievement.

Selecting a Curtiss seaplane was a wise choice as by 1911 Curtiss had made important improvements on aircraft technology of the day. For example, Alexander Graham Bell had formed the Aerial Experiment Association to encourage the development of flying and convinced Glen Curtiss to join as the engine expert. While with this group he devised and installed ailerons to ease and more carefully control turning, as opposed to the Wright brothers' use of wing warping. As Curtiss continued to build different airplane designs, he placed the elevators at the rear of the fuselage rather than in front as on the Wright Flyer, a much better aerodynamic position. It also improved the pilot's ability to see where he was going. Curtiss also invented the stepped hull that permitted large seaplanes like the Curtiss America to break water "adhesion" and take off more easily from water. It was only because of this invention that seaplanes became practical.

In 1913 a Navy search board recommended that a naval air station be established at Pensacola. Thus the city began its long association with the first US naval base devoted to the development of naval aviation and the training of pilots. Pensacola became the home base for seaplanes, dirigibles and balloons and the place where hundreds of WW I Navy pilots were trained. Leading up to and during WW II, 1,100 Navy pilots a month were trained at the Pensacola Naval Air Station and its satellite fields.

Although naval aviation had made tremendous strides since 1911, becoming an air force with many different types of carrier aircraft and a large armada of carriers, the seaplane base was still functioning when we arrived. (Flying boats eventually were phased out, the P5M being the last of the breed.) Floating in the harbor or parked on the ramp were several PBM Mariner flying boats. The PBM succeeded the legendary PBY Catalina of WW II that played such an important role in early Navy successes in the Pacific. It was much larger than the PBY, powered by twin 1600 hp R-2600 engines mounted on a gull wing. We had hoped to fly in the single-engine SNJ trainer in use at the time; however, the PBM was selected for our indoctrination.

Preparing the PBM for launch was an interesting exercise. It was not a true amphibian; on the ramp the plane was supported by large removable pneumatic wheels attached to the hull. To launch, a tractor backed the plane down the ramp into the water. Swimmers removed the wheel assemblies that included large floats so that they could be towed back to the ramp - a rather crude procedure, but effective. The plane then taxied to the pier and some five or six of us midshipmen walked on board and seated ourselves in the rear.

Once airborne, each of us was allowed to move to the cockpit for a few minutes to observe the pilots as they went through different maneuvers. It appeared pretty complicated, with lots of dials, buttons, switches and levers that the pilot and copilot watched or fiddled with as they answered questions. It wasn't too exciting, but at least we were flying. After each midshipman had his turn in the cockpit the pilots returned to land. Total time in the air was about an hour.

Completing our training at Pensacola, we packed up and got back on a train for the trip to Little Creek, home to the Atlantic Fleet's amphibious ships. The 36-hour trip seemed to take forever, as traveling to a highly anticipated destination always does. We knew that new challenges awaited, including participation in a mock invasion.

At Little Creek we were housed temporarily in barracks as we attended lectures on amphibious operations. In 1949 not much had changed from the way the Navy conducted landings that had been so successful during WW II. Troops were carried in large troopships to within a few miles of the beach, then loaded onto small landing craft for the run to shore. Larger landing craft such as LSTs would maneuver closer to shore to directly offload tanks and other vehicles. Before the troops landed the beach would have been bombarded by ships standing farther offshore and Navy aircraft operating from carriers. The Navy's responsibilities were to get the men and equipment ashore as quickly as possible and then provide continuing bombardment as requested. Once a beachhead was secured, the Navy became a supply chain for all that was needed. These were straightforward, common sense operations that had been perfected over the years, but assuring that all the pieces fit together and delivered on time was very complicated.

After three weeks of classes in which we were taught all these details we transferred to a troopship. With a small flotilla of other ships we set sail, leaving Hampton Roads behind, past the Cape Henry lighthouse and out into the Atlantic. We proceeded down the coast to North Carolina, where we would participate in an amphibious landing. We berthed in the same quarters that would have housed troops awaiting a real landing, five-high bunks in close quarters with few amenities. The good news was that we experienced the surroundings for only a few days, not the weeks that soldiers and marines endured preparing for an invasion. But it was a good lesson for those of us who might some day command such a ship where the troops onboard would be facing a bloody battle to secure a beachhead.

Off the intended landing site we drilled in manning the small landing craft that would carry us ashore. We climbed up and down cargo nets suspended over the side, wearing Mae Wests but fortunately not heavy backpacks. It was not an easy exercise with the ship rolling and pitching and the landing craft bobbing up and down. On the appointed day we scrambled down the nets and started the exercise. We circled, setting up the sequence of the waves of LCVPs that would carry us to the shore, and off we went.

At the beach the LCVPs lowered their ramps and we waded ashore. The Beachmaster, bullhorn in hand, barked directions as we ran up the beach. In the meantime larger landing craft were unloading a few tanks and vehicles. It was about as close as one could come to the real thing except that no one was shooting at us.

This ended our aviation-amphibious summer training. At Little Creek we boarded trains for New York City and Columbia, a little smarter than when we had left. We had a few days to get organized, schedule our classes, and regale our friends with wild tales of derring do.

Junior Year

At the beginning of the year I changed roommates, rooming with Jack Schleef who came from far out on Long Island and Bill Kelly from Brooklyn. Both were pre-law students, and Bill and Jack had roomed together the year before. Bill's mother, a widow, was the Congressional representative from Brooklyn. I switched roommates because Bill and I had become friends with similar interests including the Brooklyn

Dodgers. We both tried out for the baseball team, although neither of us made it, and we played on intramural teams together. Jack had been on the freshman fencing team and was on the golf team. But we definitely weren't a bunch of jocks. We were serious students; both Bill and Jack hoped to be admitted to Columbia's law school. As a threesome we successfully bid for a corner suite identical to my last one. With a flip of a coin I won the single bunk again and was usually out of the dorm before Bill and Jack were up.

After the midshipman cruise I made a major purchase. To spend weekends in Philadelphia with my new girlfriend I had to have a car. The only other way to get there was by bus or train, both expensive and time-consuming and presenting the problem of how to get around after arriving in Philly. My "Main Line" girl, now attending Bryn Mawr, wouldn't be impressed with a guy showing up on a bus and taking her out in a taxi. Gas cost only 21 cents a gallon, and I calculated that with a reasonably well running car I could make a round trip on one full tank, an important consideration in light of my financial situation.

With great sorrow I sacrificed the stamp collection that I had been adding to for ten years. Ahh, ain't love wonderful? Actually, my collection had first been my brother's, but he lost interest in the hobby and sort of gave it to me. Confiscated abandoned property might be a better description.

Through the years, despite my low income, I slowly added to the collection. Among the regular additions were first-day covers - my version. First-day covers are standard collectables issued by the post office with a made-for-the-event logo on each envelope and a short printed message. When a new stamp came out Mr. Keily, the postmaster, would save two sheets for me. I would take a block of four with serial number and glue it on an envelope with my address. Then Mr. Keily would carefully place a cancellation with date in the center of the block. I would purchase another block of four with serial number and place it inside the envelope, uncancelled, take it home and carefully store it away.

I also subscribed to a service provided by the Mystic Stamp Company in Camden, New York. They would send a small selection of stamps every month from which I picked those I wanted, paid the amount shown for each one selected, usually ten to 20 cents, and mailed the remainder back. Making small purchases and trading with a few collector friends, I added to my collection stamp by stamp. I knew each one by heart, like my little lead soldiers. It was not a large collection, but with my first-day covers I believed it had some value.

When I took it to a stamp dealer I didn't get the price I was expecting.

I had checked prices for my best stamps in a Scott's Standard Stamp Catalog, 1946 edition. Perhaps the dealer took advantage of a 19-year-old. He gave me $350, and I went hunting for a $350 used car. Somewhere stamp collectors have a series of unique envelopes with first-day issues addressed to Don Beattie, Tappan Road, West Norwood, New Jersey. (The zip code hadn't been invented yet.) I hope they have taken good care of them.

I found a 1940 Chevrolet two-door coupe, a little the worse for wear but by chance with a new paint job in Columbia blue. From 1942 to 1947 no new cars were made. Most that survived the war years had seen heavy duty, but the Chevy's tires were pretty good, the engine sounded good, and there were no metal filings in the crankcase oil. Asking price $350 - sold! From that day on I had wheels. Once junior year classes started I parked it on side streets near the campus. It turned out to be a reliable car and for two years supported many weekend dashes to Philly.

At the beginning of the year several classmates and I petitioned Columbia to establish a lightweight football team. The Ivy League had an organized program and played a schedule that included Army and Navy. Our petition was successful, and that fall we began practice as Columbia's first team. Now, as a member of the football team, I gave up my chance to be a track star. One long-suffering coach was assigned to us, or perhaps he volunteered - Mr. Carol Adams, varsity tennis coach. We were given hand-me-down equipment from the varsity and lockers in the field house. As one of the tallest and fastest on the team, I was an end. Our weigh-in limit on game days was 154 pounds. My normal weight at the time was almost 160, so I had to be careful not to eat a big breakfast or lunch before weigh-in on game day.

We played six games that first year, but the only one I distinctly recall was against Princeton on the road. It rained before and throughout the game and the field was a quagmire. The play I remember best was a kickoff. Some of us played both ways, and running down the field in the heavy rain I didn't notice that the receiver had raised his hand for a fair catch. I creamed him and received a 15-yard penalty. I do remember we lost the game. The team picture, taken at the end of the year at Baker Field in front of the Columbia Lion statue, shows 29 players.

Junior year was filled with more activities than my first two years. In addition to playing on the lightweight football team I took 19.5 credit hours each semester and held two jobs at the cafeteria and Lion's Den. But the biggest change was the frequent trips to Philly in my "new" car. I would leave after my last Friday class if a football game wasn't scheduled and return late Sunday afternoon in time for my shift at the Lion's Den. It made for a long weekend.

Merritt often accompanied me. We would take the Lincoln Tunnel to New Jersey, pick up US-1 in Jersey City and drive to Trenton, the only place in south Jersey where there was a bridge across the Delaware River. US-1 wound through small and large towns with many traffic lights, sometimes as a four-lane divided highway but usually not. By exceeding the speed limit wherever possible we could make this part of the trip in about three hours. Once across the Delaware it was a short ride to Merritt's home. The Garden State Parkway, Jersey Turnpike and the interstate highway system didn't exist; the latter would be an initiative of the future Eisenhower administration. In 1950 Ike was preparing to run for president while holding a similarly titled position at Columbia.

Third Midshipman Cruise - June 1950 to August 1950

With three years of Navy courses and two summer cruises under our belts, our third summer cruise would give us the chance to actually perform as junior officers. The title midshipman has a long history. First used by the British in the age of sail in the 17th and 18th centuries, it applied to a senior enlisted man whose assigned station was amidship at the main mast. Through the years it took on a different meaning and applied to a boy, sometimes as young as twelve, who spent three or more years at sea before receiving his commission. During the 19th century, if senior officers were injured in combat, a midshipman might be required to take command. Still later it referred to a young man in officer training at a naval college. In the US the Naval Academy at Annapolis was established to train such young men to be officers while holding the title of midshipman. Regardless of how the responsibilities changed through the centuries, on this final midshipman cruise my position in the pecking order was as the lowest ranking officer but based on a proud tradition.

For the second time I boarded a train from New York to Norfolk. On this trip, however, when the train made its scheduled stop at the 30th Street Station in Philly I planned to jump off to say goodbye to my girlfriend. As we pulled into the station, there she was standing on the platform with her mother. Eager to have as many minutes as possible with them, I jumped off while the train was still moving. What followed was another of life's most embarrassing moments. I didn't judge the train's speed correctly and tumbled down the platform, narrowly missing the steel girders that held up the roof. Skinning a few extremities and scuffing my new uniform, I walked back sheepishly for a short embrace and a self-conscious kiss while her mother looked on.

I was to spend the summer on the USS New DD 818 sailing from the Norfolk Naval Base. A new Gearing-class destroyer, she was commissioned in August 1945 three days after VJ Day. Along with three other midshipmen, also on their third cruise, I took residence on board in a junior officer's cabin. We stood watch on the bridge, in the engine room and at other stations, performing as junior officers with no holy stoning decks or chipping paint.

For the next three months, in company with other units carrying midshipmen on their summer training cruise, we participated in a number of exercises including refueling with a tanker at sea, a technique the Navy had perfected during WW II. Our ultimate destination was Quebec, where we would have our only shore leave. Surprisingly I have few specific memories of this cruise, whereas others left indelible ones.

Senior Year

With a full-time girlfriend 100 miles away, I decided I needed income additional to what I earned in the cafeteria and Sunday night at the Lion's Den. I found a night job, Monday to Thursday, at a limousine service making reservations to take our clientele to Atlantic City. The office was on Broadway five blocks south of the campus, and I walked there after dinner. I sat at a desk with a telephone and reservation book, entering a caller's name, address, and date they wished to travel. The phone rang frequently, but I still had time to finish whatever homework I brought. I closed the office at ten and went back to the dorm to finish my final homework assignments.

I had time for this job because I was no longer playing football. I had to give up being a member of the lightweight team because I couldn't make the weight. I played some intramural basketball on the NROTC team, but that was it. After I returned to the dorm Bill Kelly and I would study until about 11:30, then go to a small grocery store across the street on Amsterdam Avenue, buy milk and cookies and two different newspapers, and go back to our room to read the papers. It wasn't exactly a monkish life, but there were few frills except for weekends in Philly.

As my senior year came to a close I looked forward in great anticipation to being commissioned and serving on active duty, but with an uncertain future my Philadelphia romance began to wane. I planned to apply for aviation training at Pensacola, which would have meant continuing a long-distance relationship. And I was not ready, as were some of my classmates, to take the big step and become engaged. I don't remember our last date, but I hope it was a graceful ending to a mutually enjoyable relationship that lasted over two years. I sold the old Columbia-blue Chevy and prepared for my next life.

Midshipman plebe Beattie, backyard Norwood, 1947

Columbia College, Hamilton Hall at left, Hartley Hall center

Midshipmen Beattie and Terwilliger, 1948

Plebe midshipman cruise - USS Macon at pier in Norfolk, June 1948 (L)

Curtiss SC-1 Sea Hawk hoisted onto Macon fantail (R)

1940 Chevrolet - not the original

College yearbook photos of me, Herman Brown, Bill Kelly, Merritt Rhoad, Jack Schleef

Receiving commission on the Quad at Columbia, June 1951

Chapter III
Navy Active Duty 1951 to 1952

On June 7, 1951, in my new dress-white uniform with epaulets on the shoulders, I was commissioned Ensign USN at a ceremony held on the Quad at Columbia. I had requested flight training but was denied. I learned that newly commissioned officers were required to spend one year in the fleet before being eligible. I was ordered to report to the USS Liddle, APD-60, home-ported with the Atlantic Fleet's amphibious forces at Little Creek, Virginia. However, in June the Liddle was serving a tour in the Mediterranean with the Sixth Fleet. I was ordered to report to the Navy Bureau of Personnel (BuPers) in Washington, D.C., to await transportation. In the meantime I was granted ten days' leave that I spent with my parents in their new home in Fair Haven, New Jersey. I borrowed $200 to buy a complete set of uniforms and had a gold stripe sewn on the sleeves of my new dress blues.

Reporting to USS Liddle

Finishing my leave I met classmate Doc Sylvester, who had similar orders but to a different ship. We took the train to Washington and reported in. While awaiting transportation we were assigned to review NROTC candidate files at a BuPers office located in a temporary building in Anacostia, Virginia. As we were not assigned official Navy quarters we tried to find suitable accommodations for the next two or three weeks, but nothing was available that we could afford.

I played my trump card and called my Columbia roommate Bill Kelly's mother Edna, still the Congressional representative from Brooklyn. She arranged for us to stay, at a greatly reduced rate, at the Hay Adams hotel where she lived when Congress was in session. What a deal! The Hay Adams at that time was one of the most luxurious hotels in Washington; perhaps it still is. I also asked her to use her influence to change my orders and allow me to report immediately for flight training. She too was turned down by the Navy brass; the letter she received (which I have) politely suggested I report to the Liddle as ordered. Oh, well, no harm in trying.

Doc and I commuted daily to Anacostia on the trolleys that ran on many routes from downtown to the suburbs. I don't recall exactly what we did with the NROTC files, but I assume it was something useful. The hours weren't long, so we had time to take in the Washington night life. There were many "government girls" working at offices all over town; we met a few and generally enjoyed ourselves. This turned out to be bad news for Doc. Right after graduation he had become engaged, and he was so busy running around with me that he forgot to keep his fiancée up to date. He received several angry messages from the lady asking why he wasn't calling her every day. I assume he redeemed himself but am not sure what excuses he gave.

My transportation became available on July 6. I flew on an R4D from Navy Anacostia to NAS Quonset Point, Rhode Island, then on a R5D to Gander in Newfoundland, across the Atlantic with some 20 other sleepy passengers to refuel at the Canary Islands, and finally to Port Lyautey, Morocco. It was a 24-hour flight sitting on a thin canvas seat secured along the fuselage with our gear tied down in front of us, very hard on the sitter, noisy, and impossible to take more than a few short catnaps.

Transportation to Nice, France, my final destination, was not available at Port Lyautey for three days, so I had a chance to do some sightseeing in the Moroccan bazaars. I watched camels carrying cargo and passengers and bought a camel skin wallet. The eventual flight to Nice was uneventful.

On July 12 I joined the Liddle, commanded by LCDR. Kenneth W. Miller, anchored off the French Riviera at Golfe Juan, near Cannes. As the most junior officer I was assigned a variety of duties, some cast off by more senior shipmates: Assistant First Lieutenant, Assistant Gunnery Officer, Boat and Morale Officer, as well as several boards including cryptography and mail censor. Most of my waking hours were taken up with paperwork, a rude introduction to the working Navy. My assignment as Assistant Gunnery Officer was interesting as the Liddle had no listed Gunnery Officer.

Recently, while reviewing the ship's logs that I ordered to confirm some of the forthcoming dates, I came across an error in the paperwork. The ship's log for July 12, 1951, does not show that I reported onboard. This is a strange omission, as it is standard procedure to record new crew reporting or leaving, not only officers but also enlisted men. The log for July 12 did record that seaman W. W. Brown was processed for discharge. Perhaps the officer of the day when I arrived, LTJG Tom Urich, who had been standing watch continuously for the last twelve hours, was too tired to write it down. Or perhaps I made up the following stories, never reported, and enjoyed life in Cannes! Three ensigns that reported for duty at later dates were logged onboard.

USS Liddle - APD-60

APDs are destroyer escorts (DEs) converted during WW II to carry Underwater Demolition Teams (UDT), now called Seals. Before an invasion UDT swimmers were first ashore to remove beach obstacles or to conduct special operations. These activities required APDs to lead the invasion, dropping off swimmers as close to shore as possible. APDs had a draft of a little over thirteen feet, which allowed them to maneuver close to shore. As a class they had a displacement of 1,400 tons. Top speed was 24 knots. The Liddle had one 5"-38 turret forward and several 20 and 40 mm antiaircraft guns. On the fantail were depth charge racks.

The APD's major difference from a basic DE was the addition of four landing craft (LCPRs) suspended on davits amidship, two on each side. Each LCPR weighed about seven tons, and the four suspended high above the waterline made the ship top-heavy. Even in light seas she might roll precariously. In heavy seas one wondered if she would right herself after a deep roll. It was easy to identify APDs when docked with other ships; they all had a permanent six-degree list to port and the tilted masts made them easy to find. Why the list? I don't recall ever hearing a good answer.

Normal crew size, according to official Navy data, was 14 officers and 150 enlisted men. That must have been a full wartime complement. When I reported I became the ninth officer onboard. At one time I am sure I knew what the crew complement was, but I don't believe we had 150. In addition to the Captain there were Executive Officer, LT John Bigham, Operations Officer LT Bill Eason, Assistant Operations Officer LTJG Tom Urich, First LT Mel Taylor, Supply Officer LTJG Fred Simcich, Chief Engineer LTJG Fred Hoffman, Assistant Damage Control Officer LTJG Bill Myers and me. Taylor, my immediate boss, was a mustang, a name given to an officer who had worked his way up from enlisted man. Several of the officers had been recalled for duty as the Korean conflict continued. In addition we had two Underwater Demolition Team (UDT) officers onboard with their team. At times we also carried small US Marine detachments.

Officers' mess and quarters were under the supervision of black enlisted personnel who, among other duties, served our meals in the small wardroom and kept the coffeepot full. As the new Morale Officer I quickly introduced myself to the crew and recruited a basketball team. We had some good athletes among the enlisted men and we played other ships when we were ashore. But after more than 50 years I don't recall the names of any of the chiefs or other enlisted men; my apologies to all.

A lesson never covered in our Naval Science courses at Columbia was how to interact with the crew without seeming to be either their buddy or an overbearing stuffed shirt. How could I balance the roles of an officer they were required to salute and someone they would willingly follow and enjoy being with in less formal circumstances such as team sports? I'm not sure I ever achieved that balance. To some it must come naturally, but I found it to be a difficult transition from college where all my friends and acquaintances were equals. Chiefs and petty officers obviously knew more than I about how to carry out specific duties, yet as the officer in charge of a work detail or duty station even a young ensign was expected to supervise. In some respects it was easier to work with chiefs, as they could be treated almost as equals. Chiefs had been around long enough to know how to keep the relationship professional. I struggled with this problem as the months went by, especially when working with new recruits.

The Liddle, I was to learn, had a fascinating history. Named after a U.S. Marine killed on Guadalcanal, she was commissioned in 1943 and had participated in many Pacific campaigns. For her new life she had been taken out of mothballs at Green Cove Springs, Florida, in October 1950 to augment the rapidly expanding Navy seeing action in Korea. But it was her combat role in the Pacific during WW II that made her famous.

After taking part in several earlier invasions and then landing troops at Ormac Bay, Leyte, on December 7, 1944 (how about that date?), she came under kamikaze attack. Her crew was credited with shooting down five planes, but a sixth crashed into the bridge, killing most of the officers and many crew. The damage and casualties resulted in a ship with no one in command, and she began to circle helplessly. Help came from nearby ships and the repair crew from the USS Cofer APD-62 finally brought her under control. The Liddle sailed back to San Francisco, was repaired, and rejoined her unit to participate in the final Philippines campaign. An interesting part of this story is that, after moving to Florida in 1996, I learned that one of the members of the Cofer's repair party, James R. Snellen, lived just down the street. He has written a book of his experiences for those who may be interested in the details of the Liddle's rescue.

Four days after I reported we weighed anchor and set sail for Sardinia. Once under way, as the most junior officer, I was assigned the least favorable bridge watch, the mid-watch from midnight to 4 AM. With a standard four on and eight off duty cycle my next watch was from noon to 4 PM. On the mid-watch I was more an apprentice than an officer, observing the OOD, taking bearings, watching the radar screen, and keeping the ship's log. My noon watch was much more interesting as the skipper soon allowed me to con the ship during maneuvers such as reorienting the task force screen during a course change. This required using a maneuvering board, good judgment, and a mental picture of what all the ships in the task force would be doing.

When we operated as a unit of the Sixth Fleet we were often the point on the screen, steaming ahead of the task force. Destroyers or other DEs formed the rest of the screen, sonars pinging away to detect submarines. This was the height of the "Cold War," and we knew that Soviet submarines were always shadowing our task force. By the first week in August I qualified as junior officer of the deck (JOOD) and stood the noon to 4 PM watch with no other officer present. My instructions were to call the captain or exec if there was any unusual activity!

To complete my story of standing bridge watch, I soon stood the midwatch alone. At the first opportunity ashore I bought a pipe. I was not a smoker, but my reasons for making this change were twofold. First, I thought it would give me an air of authority: officer of the deck standing on the bridge, pipe in mouth, staring off across the dark sea and keeping all his shipmates safe. I had probably seen such heroic poses in WWII movies. Second, it takes great skill and perseverance to keep a pipe lit, especially for a novice. Thus, I would spend time adding tobacco and trying to keep it burning while not inhaling the smoke. Watch time flew by! There were some unpleasant consequences, however. Residue from the smoke mixed with saliva burned my tongue, and there was a lingering sour taste in my mouth. What price glory? I confined my pipe smoking to time on watch.

UDT-2, based in Little Creek, furnished teams to several ships. Our team consisted of two officers and 24 enlisted men. Its commanding officer was LTJG Boule; second in command was a young ensign, Edward M. "Toad" Swanson. Although attached to the Liddle, the unit had its own operating procedures and at times dictated what we would do. As the two most junior ensigns onboard, Toad and I bunked together in a communal compartment below the fo'c'sle that could hold four officers.

Toad got his nickname from his swarthy complexion and heavily pockmarked face, the result, I assumed, of a bout of childhood chicken pox or acne. He didn't seem to mind our joking with the name. Regardless of his looks, he was a friendly fellow, a Georgia Tech graduate with a soft southern drawl. He would entertain me with stories of UDT training that was extremely difficult and dangerous. UDT recruited really tough guys. You wouldn't want to make one angry.

In order to stay fit the UDT team was required to log time swimming and conducting drills in the water. One of the Liddle's responsibilities was to assist in this training. The team would usually have one or more swimming exercises each week. We would slow, and the swimmers would jump off the stern. We would then steam away for the next two hours or so, then circle back to pick them up, a good drill for our navigator.

Getting them back onboard was more complicated then dropping them off. We would slow and lower one of the landing craft with an inflatable rubber raft tied to its port side. Preparing for their pick up, the swimmers had lined up in the water with about 25 yards separation between each other. The biggest guy on the team was the snatcher, and he would be the first picked up. He would then lie in the rubber raft; as we ran along the line of swimmers each would raise his right arm which the snatcher would grab and quickly roll him into the raft. Once in the raft he would then jump into the LCPR before we reached the next swimmer. Down the line we went without stopping, and in a few minutes all the swimmers were in the boat. Then we would hoist the LCPR on board. All was done with great precision and with no injuries that I recall. The team was always ready for any assignment. In my duty as Boat Officer I participated in the UDT operations using one or more of the four landing craft. Although deadly serious training, it was great fun. The close camaraderie among the team permitted continuous joking and pranks.

From Sardinia we sailed to Phaleron Bay, Greece, the port serving Athens. LCDR Robert J. Sammons reported onboard to relieve Captain Miller. For Sammons, a tall, lanky Annapolis graduate, the Liddle would be his first command. He had been a member of one of the abbreviated Naval Academy classes during WW II that graduated after only three years as the Navy speeded up commissioning officers to man the vast armada of ships fighting a two-ocean war. He seemed like a nice fellow and, having established a good relationship with LCDR Miller, I looked forward to his leadership. While in Greece a few of us hired an old Cadillac limo and driver and took a tour of places far from Athens that most tourists probably never see.

Change of Command

We upped anchor and steamed toward Souda Bay, Crete, staying with a small task force for four days, then broke off to begin independent operations. On August 1, while underway, Miller handed over command to Sammons in a formal ceremony.

The next day we maneuvered to begin a complicated UDT exercise to test security of units of the Sixth Fleet at anchor in Souda Bay. To do this we simulated a night commando attack. The ships had been forewarned of the exercise but not of the time it would occur or the type of attack. The sentries, we hoped, would not be carrying loaded guns.

Souda Bay, on the north shore of Crete, was a large anchorage protected by a long peninsula on the seaward side, leaving a rather narrow entrance into the harbor. The first line of security was a small picket boat that patrolled the entrance. Just before dawn, a few miles outside the harbor, we lowered an LCPR, proceeded to the harbor mouth, and captured the picket boat. A few of the Liddle crew and I stayed with the picket boat and continued the patrol in case someone inside the harbor was watching. Our UDT force then swam into the harbor and simulated an attack on the anchored ships. Several swimmers arrived undetected at the flagship, the USS Sierra AD-18, shinnied up the anchor chain and captured the Marine guard and the Admiral.

We later learned that the Admiral was upset at how easily this was accomplished. New security directives were issued, which, of course, was the object of the simulation. An example of a change put in place: now, when a Navy ship is moored or anchored, it deploys floodlights at the waterline so that swimmers attempting at night to get close can be seen by the deck sentries.

After our simulated attack we came back onboard, and the Liddle entered the harbor and anchored. Miller was detached, and during the next days we were involved in amphibious operations at different beaches along the coast. Each night we would anchor either at Souda Bay or another protected harbor. Our UDT team participated in each exercise, clearing beach obstacles, and we had lots of practice lowering and raising our LCPRs.

Ten days after our simulated attack we hoisted anchor and left Souda Bay en route to Malta. The next day we ran into rough seas and the Liddle began to roll heavily. A storm was brewing in more ways than one. By now I was accustomed to the rolling and had no problem. We all went about our business. Lunch was announced. If the ship was rolling heavily when meals were served, the wardroom stewards would install "fiddle boards" at each place setting. They consisted of low raised dividers that sectioned off an area in front of each diner to keep plates from sliding around on the table. As was the custom, we entered the wardroom first and sat down. Our new skipper soon appeared. We all stood and after he sat down resumed sitting at his invitation.

First course was soup; conversation around the table was muted compared to the usual joking and small talk as we waited for Sammons to lead the discussion. The next course was roast pork, potatoes and a vegetable. We noticed our new captain staring down at his plate and saying nothing. Suddenly he jumped up and dashed out of the wardroom. Where he went we couldn't be sure, but we assumed it was to the rail to lose the first course. He didn't return; with rolling eyes and in silence we finished the meal. This was not exactly how a new commander wanted to impress his officers and, unfortunately, it presaged difficult times ahead. Sammons turned out to be a much different commanding officer from the one we had become accustomed to.

Life onboard a Navy ship might be likened to living in a parallel universe or newly discovered solar system or galaxy. The captain is the "black hole" or bright star around which everything revolves. Although shipboard life is governed by Navy regulations and tradition, a captain has great latitude in interpreting them. Even when operating as part of a large task force, the flagship commander is a disembodied voice occasionally coming over the radio: "Course change to 090. Execute." The execution is up to the captain. But we frequently operated independently of the main force. Our captain's actions were unobserved except by the ship's crew. Sammons' style of command in the months that followed began to raise serious questions.

Despite onboard concerns, liberty was frequent. After our Malta UDT exercise where we "captured" the radar station we visited Marseille. As the smallest ship in the task force we were invited to tie up in Port Vieux, the small harbor that connected the city directly to the Mediterranean. The rest of the task force anchored offshore or moored at piers far from the city. We put our brow over the stern onto the main plaza and La Rue Canebiere. Sightseers and pretty women were just 20 feet away. It was an outstanding but short visit with the hospitable French! While moored, Fred Simcich hired some local workers to paint the ship's hull with brushes on long poles as they stood in skiffs tied alongside. In two days we looked very shipshape.

Applying for Flight Training - Liberty in Naples

From Marseille, at the end of August, we sailed back to Cannes. While there I learned that the one-year fleet service requirement before being eligible for flight training had been lifted. I requested permission to take a flight physical on the aircraft carrier USS Oriskany anchored nearby. Flight physicals had to be administered by a flight surgeon, and one was on the Oriskany. When Toad heard of my intentions he made the same request, and orders were cut for both of us. We took a small boat over and I passed the physical. I don't recall Toad's result but assume he passed as he was in great physical condition. Then began a long wait. The physical was just the first step, and I hoped that the war in Korea required many new pilots.

From Cannes we sailed to Naples, conducting various exercises. En route, under my command, we fired one round of the 5" for the first time since I had reported. Liberty in Naples was especially memorable. Fred Simcich had relatives living there. They met us at the pier and gave us a grand tour of the city and surroundings, ending at a lovely outdoor restaurant high on a hill overlooking the harbor. The owner treated us to a wonderful five-course dinner. When we had finished he pulled a large gold piece from his pocket shaped, to the best of my recollection, like a pepper. He insisted we all rub it for good luck and to bring us back to Naples. This may have been a lucky talisman dating from Roman times. My rubbing was successful; I had good luck and returned to Naples in 1954 when my squadron was on the aircraft carrier USS Mindoro.

One of Fred's relatives owned the Borselino hat factory. Borselino hats were, at the time, top of the line; in movies and press photographs from the '40s and '50s most men are seen wearing fedoras. I bought one for my father which came tightly rolled in a tube. When I gave it to him a few months later it popped out into its original shape without a wrinkle! I also bought a lovely cameo with a delicately carved classic head of a woman for my mother.

Returning to Little Creek, Virginia

From Naples we sailed west to Oran, Algeria, where we transferred all explosive ordnance used by the UDT plus the 20 and 40 mm rounds and 5" shells to the USS Cobb APD-106. The Cobb had its own UDT detachment on board. With that our tour with the Sixth Fleet ended. We set sail from Oran on September 20, in company with a small task force until two days out from Norfolk. Then we proceeded independently, arriving at the Convoy Escort Piers on October 1. The UDT detachment left the ship and returned to Little Creek. Many of our crew and officers went on leave, while those of us left behind formed a skeleton crew and kept the ship running. This was my first time back at Norfolk since my final midshipman summer cruise in 1950. Now my duties were much different. On October 8 we left the Convoy Escort Piers and moored at Little Creek, the Liddle's home port.

Standing watch in port, or off watch, meant continued formal training with written tests to demonstrate that I was familiar with all aspects of ship equipment and procedures. These included working with the engineering department to understand the ship's propulsion, power and mechanical systems in preparation to qualify as Officer of the Deck (OOD), the ultimate goal of a ship's officer. Interestingly, to qualify as OOD in port was more difficult than when underway. An OOD in port might be responsible, in an emergency, to get the ship underway with minimum crew if senior officers, including the skipper, were not available.

I recruited crew members for a volleyball team as well as the basketball team and we competed in formal leagues. We did very well, usually playing against teams from much larger ships. Buying uniforms with the ship's name was good for morale, and some of the crew would come along to cheer. Taking the teams to different gyms was lots of fun. Sometimes we played against teams in the well-deck of their LSTs. As our basketball team was at the top of the league, I convinced Sammons to let us stay ashore for a tournament while the ship made a short deployment.

Bachelor Tom Urich had left a 1948 Plymouth sedan at Little Creek when the Liddle sailed in May to join the Sixth Fleet, so we had transportation when off duty. We began a limited social life, dating nurses at the Portsmouth Navy Hospital. It was nothing serious; we weren't around long enough to keep ladies waiting by the telephone. Nevertheless, I recall dancing at officers' clubs with a lovely sweet-smelling partner while a Tony Bennett record played:

"She wore blue velvet, bluer than velvet was the night,

softer than satin was the light from the stars."

"Hormones" were aching to be relieved. But in mid-twentieth century America members of the opposite sex were much more reserved than what I have observed in their later day counterparts, and my ardor went unsatisfied. I don't recall my dancing partner's name or how many dates we had, but they were definitely not enough.

We started a Friday night sing-along at the Little Creek officers' club, sometimes with and sometimes without piano accompaniment. Eventually we printed a few sheets with the words of favorite songs for those who joined our group. Returning to the club many years later while attending a VS-30 reunion at the Naval Air Station, I was surprised to find that Friday song time had become an established tradition with a multi-page song book and full-time pianist. The song book gave no credit to the USS Liddle officers who started it all in 1951.

Operational Readiness Inspection

Training never ended. We left port twice to conduct exercises off the Virginia Capes with two submarines, USS Sea Lion SS-315 and USS Runner SS-476. During these exercises we also conducted gunnery practice with a towed target and fired five rounds from the 5" in local control. We were scheduled for an Operational Readiness Inspection (ORI) in November. As this would be his first ORI as a CO, Sammons was very tense as the results would determine his fitness to command and to be selected for promotion to ever more important assignments. We drilled when we were underway, practicing some of the types of exercises that would be scored.

An ORI is the most complete review of a ship's ability to carry out its duties. On the appointed day inspectors came onboard with experience in all the ship's functions. Thus, for example, engine room inspectors were in the engine room as different drills and emergencies were conducted. Similarly, others were on the bridge and graded how Sammons got underway from the pier. Some inspectors stayed on the bridge to observe all our maneuvers as we left Hampton Roads and graded the crew's response to emergencies such as man overboard. However, most of the drills were held off the Virginia Capes and included firing our guns.

As the Assistant Gunnery Officer (we still didn't have an assigned gunnery officer onboard) my primary responsibility was to oversee firing all of our guns. This would be very interesting because, since I had reported onboard, we had fired the 5" twice, one round in the Med and the five rounds described above. The 20 mm was fired once when we exploded a drifting mine in the Mediterranean and again, along with the 40 mm, during the above practices. Firing the 20 and 40 mm guns in the ORI went off without a hitch thanks to training before I reported and the ability of the gun crews, not due to my direction.

The next evolution was to fire the 5" at a sled pulled by a tug a few thousand yards off to port. The first firing was in local control. That meant I stood at the rear of the turret with my helmeted head stuck through a small hatch and, with headphones and mike, gave the crew bearing and distance numbers that they used to train the gun. We fired the first round; I observed the splash and corrected the bearing and distance. We fired eleven more rounds in the same manner, never hitting the sled but with some near misses.

The next exercise to be graded was firing the gun under automatic control. This required me to man the small fire control director mounted above and aft of the bridge, the first time I had done so for active firing. I was unconcerned; it was a straightforward operation. With helmet, mike and earphones on I was in voice contact with the crew in the turret. The first procedure was to match pointers for bearing and elevation on my controller with those in the turret. Initial matching was done with the turret pointed directly forward, dials in the turret and at my position reading zero. When the gun crew in the turret reported pointers matched, I slewed the director to port to pick up the sled. The turret slewed to starboard. End of exercise with one very exercised CO.

We returned to Little Creek and the skipper confined me to the ship until the problem was fixed. The gunnery chief and I worked the next 24 hours without a break. Using very old and faded wiring diagrams we traced every wire from the fire control director down through the bridge, under the main deck, and back up into the turret. As expected, we found that some wires had been crossed and the turret, not knowing any better, had precisely followed commands. The problem was solved, but the captain was still incensed. I don't recall what grade we received on the ORI, but obviously we had failed a major component of his first ORI when in command.

Sammons blamed me even though I had no way of knowing about the problem beforehand. Putting the turret and fire control director through its paces before the ORI, without firing, would have found the problem, but it was above my pay grade to suggest that this was needed. I assumed it had been tested before I reported. My guess is that the mixup had occurred when the ship was mothballed or taken out of mothballs.

Sammons never forgave me, and we had a difficult relationship from that day forward. I also believe he was peeved that I had asked for flight training soon after he took command. Perhaps he thought it reflected badly on his leadership. My fitness report of December 1951 reflected his opinions. He accused me of having a "short-timer's attitude." There was probably some truth to this observation, as I was waiting anxiously for the orders that would send me to Pensacola. However, I always felt that I carried out all my duties in a professional manner. Despite my "short-timer" attitude he noted that I was the "fastest Ensign in progress toward qualifying as officer of the deck." He also indicated he would be pleased to have me as an officer under his command and checked two areas on the fitness form as outstanding. I must have been doing some things right.

Moored in Hampton Roads

On our return to Norfolk after another exercise we learned that, rather than berthing at Little Creek, we would tie up in Hampton Roads, moored to a huge buoy in a nest of destroyers. Little Creek was undergoing some major changes, enlarging the piers so that the larger amphibious ships such as LSTs and LSDs could tie up closer to base facilities.

We moored to Buoy How 30 in the middle of Hampton Roads. There were usually three or four destroyers in the nest with us, all moored to the same buoy. Gangplanks were laid from ship to ship and, as one or another ship would leave or join the nest, our position would change. Sometimes we would be on the outside, sometimes in the middle. When not the outside ship we had to cross over to the destroyer on the outside to reach a liberty boat to take us ashore, each time saluting the OOD and asking permission to come aboard.

Hampton Roads in November and December of 1951 was like being in the middle of the north Atlantic; it was very cold and at times the water was rough. At least it seemed that way. A strong wind usually whistled in unimpeded from the west. Being on watch was the coldest I ever remember. Our ship had nothing resembling central heating, so many of the compartments were not much warmer than the outside temperature. When inside we were at least out of the wind.

The only good news was that the duty of checking the mooring lines between ships and to the buoy was alternated among the ships in the nest. This extra duty did not come up very often. At night it meant walking around with a flashlight, peering over the sides or bows of all the ships to test the many lines. Occasionally an exciting event livened the day. One afternoon, while the destroyer USS Robinson DD 562 was maneuvering to come alongside and join our nest, she rammed us, hooked our anchor and caused some damage to the hull. But most days we just hunched up and hoped our stay in purgatory would soon end. As the wind direction shifted our position relative to the buoy would also swing around, thus the watch on the buoy mooring was very important. One could visualize the lines breaking at night and four or five destroyers drifting helplessly down the bay to collide with whatever they ran into. We succeeded in averting such a catastrophe.

We thought twice about going ashore on liberty in a small open boat, but staying on board wasn't very appealing either. Usually we arrived at the pier colder than when we started, having been splashed a few times with cold water. But by this time our basketball team had entered into league play. In spite of the discomfort we frequently had to make the trip for practice and games. For team members this was a welcome change from staying onboard a cold ship. Finally, after 28 days, we received a reprieve and on December 14 hauled in our mooring lines, left Buoy How 30, and returned to Little Creek.

After the ORI problem, and continuing for the rest of the year, Sammons' behavior became more and more erratic. We went about our duties not knowing when he would berate someone for what we thought was a normal activity. The wardroom became an uncomfortable gathering place as the situation worsened, and we tended to avoid it except for meals. From Sammons' first day on board we noticed that he let his fingernails grow very long and, to make matters worse, didn't clean them. We could not avoid seeing this when sitting with him at meals. An indicator of some behavioral disorder? Not being a psychiatrist, I can only guess. In the meantime Bill Eason, an easygoing Louisianan, had been promoted to Executive Officer, and from time to time we quietly discussed Sammons' behavior with him. He always cautioned patience; "Let's see how events play out before I take any action."

Escorting LSS(L)s Bound for Korea

On the day after Christmas we left Little Creek and proceeded to Green Cove Springs, Florida, to escort two amphibious landing craft, LSS(L)s 57 and 120. They had just come out of mothballs and were destined to sail to Panama via Charleston, South Carolina. I don't recall why we took this circuitous route, but it was probably to outfit the LSS(L)s after being in mothballs. They had a long voyage ahead, for after leaving Charleston they would transit the Panama Canal and then sail across the Pacific to participate in amphibious operations in Korea. LSS(L)s were classified as oceangoing amphibious support ships.

After being moored for eight days at Charleston we rendezvoused with the LSS(L)s outside the harbor just as a strong nor'easter was approaching. For the next few days we suffered from the high waves and swells it spawned. We were uncomfortable rolling from side to side in heavy seas, but as we watched the progress of the LSS(L)s, with waves frequently breaking over their bows, we knew their small crews were really suffering.

At just over 150 feet and a few hundred tons, LSS(L)s were hardly oceangoing vessels as they were flat-bottomed with very shallow drafts. In such sea conditions they barely made ten knots. At one point we had to lower one of our LCPRs in heavy seas to transfer an injured seaman from one of them to the Liddle for medical treatment. I don't remember what he needed, but it couldn't have been too serious as we did not have a doctor onboard. Perhaps, with all the rolling and heaving, his CO thought he needed a more stable place to recuperate.

To make matters worse, we had following seas for three days that made steering difficult at the slow speed we maintained to stay with the LSS(L)s, at times only five knots. Wallowing along, we eventually chose to steam back and forth at higher speed to reduce our rolling and pitching, reversing course every three or four hours. The following wind created another problem. On the southward course to Panama stack gas from the funnel enveloped the bridge. Coming off watch the first thing we did was take a Navy shower to get rid of the smell and grime.

Thankfully, after eight days we reached the Naval Station, Balboa, Canal Zone, and bade our charges goodbye. Both our crew and theirs were happy to stand on terra firma again, if only for a few days. It was warm and sunny, and we were able to do a little sightseeing before going back to sea. On January 18 we were underway again and five days later arrived back at the Convoy Escort Piers at Norfolk Naval Base. Our return trip was much less stressful than our voyage south. Awaiting us was Lt. Robert Donovan, our newly assigned Gunnery Officer. The CO would no longer have to depend on an ensign to make his 5" turret function correctly.

Skipper Relieved - Orders for Flight Training

In February Sammons' erratic behavior came to a head. I don't recall the exact incident that triggered a reaction, but Eason agreed that our CO needed help and reported his condition to superiors. On February 18, after less than eight months, he was relieved of command and transferred to the U.S. Naval Hospital at Portsmouth, Virginia. By coincidence, a book we were unaware of, *The Caine Mutiny*, was published at almost the same time. It describes the fictional CO of a destroyer minesweeper with serious behavioral problems. One of the main characters in the story, an ensign, had been a midshipman at Columbia. Adding to the coincidence the author, Herman Wouk, was a Columbia graduate and had served on a destroyer minesweeper about the same size as the Liddle. Perhaps Wouk encountered a situation similar to ours or considered such a problem possible after serving on a small ship. Regardless, the result was his Pulitzer Prize-winning novel. When one's life is confined to a world 306' long and 36' 10" wide, the captain's behavior determines whether or not that world is a reasonably pleasant place to be, notwithstanding that Navy crews follow old traditions and are subject to strict discipline in all they do.

To my great relief, my orders to report for flight training arrived at the end of February 1952. I was directed to be detached from duty on the Liddle and report to Pensacola in March. By this time we had a new captain, LCDR C. E. Hart, III. He signed my orders for the change of duty. Now it was up to me to live a long-time dream and qualify as a naval aviator.

USS Liddle - APD-60

Recovering UDT swimmers.
(Photo courtesy of the National Navy UDT-Seal Museum, Fort Pierce, FL)

Liberty in Greece with Bill Eason, "Toad" Swanson and Tom Urich

The Liddle's basketball team and coach eating watermelon, Crete 1951

Chapter IV
Flight Training 1952 to 1953

Pensacola

Saying goodbye to shipmates with whom I had shared so many adventures, good and bad, in just eight short months was a little sad but I don't recall a farewell party. From Norfolk I flew to Pensacola, arriving on March 26. As I was driven through the base in a Navy bus, it didn't appear much had changed in the three years since my second midshipman cruise. However, the Naval Air Station was humming as the Korean war increased the demand for pilots. Reporting in, I was assigned to officers' quarters in a wooden barrack-type building, undoubtedly of WW II vintage, but I had my own room. There was a common area in the front that included a few chairs and a pool table. Meals were taken in a large officers' mess hall a short distance away.

Pre-Flight Training

I was enrolled in a Pre-Flight Officer Indoctrination Course with 24 other officers of various ranks. It was an unusual class, the first of its kind, because a short time earlier the Navy had decided to decommission its blimp squadrons and give those pilots a chance to qualify for heavier-than-air. In our class were two CDR's, a LCDR, and a few full LTs, all much older than the rest of the class. But we all got along fine and dispensed with the usual courtesies afforded senior officers. After all, we were really in competition with each other as there was the chance that not all would go on if grades and instructor appraisals weren't satisfactory. It may have been difficult for our instructors, some of whom were junior officers and chiefs used to instructing cadets, but I don't recall any problems. It was a great bunch of guys, and I went all the way through training with several until we earned our wings. One of those classmates, Frank Garrard, was eventually assigned to the same active duty squadron as I and we were BOQ roommates at Norfolk when assigned to our first fleet squadron.

Pre-flight training lasted six weeks. It consisted of courses similar to those I attended during my summer cruise but with much more detail and graded exams. Cadets were also included in some of the classes. Aviation cadets usually had two years of college before they were eligible to apply for flight training and, of course, were younger, especially compared to the senior officers in my pre-flight class. According to my fitness report, I ended up standing 12 in the class of 25.

The culmination of pre-flight was survival training. One of the final tests was the "Dilbert Dunker." In his flight suit the trainee was strapped into a crude mockup of a cockpit. The mockup was on rails about ten feet above the large pool where the swimming tests were held. When released it slid down the rails and turned over when it hit the water, leaving its occupant strapped in upside down as it sank. To make the test more difficult, a thick piece of wood was placed where a pilot would normally be sitting on his parachute, adding a little buoyancy to the test to keep him stuck upside down in the mockup when it turned over.

Before starting we were told to exit the mockup on a specific side to test whether or not we could think clearly in such a predicament. On reflection, this test was an early version of "water boarding." We watched movies of how to pass the test that included actual footage of Navy planes forced to land in the water instead of back on the carrier. In most cases, after hitting the water, the planes flipped over and sank.

Down I went holding my breath. The mockup turned over, I unstrapped and tried to swim out. But with the board that I had been sitting on now trying to float to the surface, and the air trapped in my flight suit holding me against the seat in a head-down position, I had a problem. I could see the "rescue" divers starting to swim over to me, and in a last effort before running out of air I swam free and got clear. This wasn't bad considering I could hardly swim a stroke. I still hadn't learned from my Boy Scout days. But in my haste I had swum out on the wrong side! I climbed out of the pool coughing up water, climbed the stairs to the top of the rails, and was strapped in again. Down I went. Practice makes perfect and I easily swam out to the side I had been instructed.

The swimming test, about 200 yards in length, was a problem, and not only because I wasn't a swimmer. A few days earlier I had been at the beach with classmates playing "horse" in the surf. When Frank Garrard, my rider, jumped on my shoulders and clamped his legs around my neck, he forced water under pressure into my right ear and I felt something pop. It was painful and required a trip to sickbay, where I learned that my eardrum had been punctured and now had a small hole. I could not be cleared to take the swimming test. Instead, I faced a two-week recuperation after which, if all went well, the hole would close normally. If it didn't close, I would not be allowed to continue flight training. Now that was a real bummer! I didn't want to think I might be assigned to another ship like the Liddle.

I had to say goodbye to my new friends who had completed pre-flight and were being transferred to Whiting Field to begin flying. But the hole closed normally and I passed the swimming test doing a backstroke. That was the good news. The bad news was that I was now two weeks behind my pre-flight class. I packed up and went to Whiting Field. On arrival my quarters and messing were similar to those I had in pre-flight, but now I went to ground school classes with cadets.

Whiting Field - Soloing

Whiting Field, about 25 miles northeast of Pensacola near the town of Milton, has an important place in naval aviation. It is where thousands of Navy pilots learned to fly and made their first solo flights. Soloing was an important first step in becoming a Navy pilot. Each of the next steps would be more difficult. Not counting ground school, the Whiting flight syllabus was composed of 19 dual flights with an instructor; the twentieth would be my first solo. I can relive them to some extent because they were all carefully recorded in my official Aviators Flight Log Book which, of course, I still have.

At Whiting there was one big difference from living at Pensacola. At 6 AM every morning plane captains, enlisted men, would start the engines of all 100 or so planes on the line. We didn't need an alarm clock to get out of bed on time. The plane captains would rev the engines to full power to make sure the planes were ready and then top off the fuel tanks. Before going out to the plane assigned we would check the "yellow sheets" to see if any problems had been noted from days past so as to be ready in case the same problem appeared again. Before getting in the cockpit we completed a "by the book" ground check of the plane, monitored by the instructor, to be satisfied it was ready to fly. On completing a flight we filled out the "yellow sheet" form noting any discrepancies. The plane's serial number, flight duration, and the name of the instructor pilot who was in the plane with me were carefully entered in my log book.

My instructor, who had been recalled at the beginning of the Korean war, was Lt. Al Wistle. He had an easygoing attitude even though he was returning to the Navy from Seattle where he had been living a peaceful civilian life. Always unflappable, he was probably the main reason I progressed with few problems. He was in the back seat during all flights except for check rides, when another pilot evaluated my progress. Check flights were given on flights 12 and 19, the last to confirm I was ready to solo. If the check pilot didn't give me an "up," I would have to repeat a flight with my instructor covering whichever procedure the check pilot didn't like and repeat the check flight.

We flew the two-seat North American SNJ Texan that had a long history as a Navy trainer. The version we flew, the SNJ-6, began flying in 1944. It was powered by a Pratt and Whitney 550 HP radial engine that gave it a top speed of about 200 mph. It had a variable pitch propellor, retractable wheels and flaps.

Starting with our first flight we followed Navy procedures as if we were operating from a carrier. Immediately after takeoff wheels up and then flaps, and we made a slight left turn to clear the field. On landing, we broke over the upwind end of the field, making a tight left turn, and put wheels and flaps down as we turned to go downwind maintaining a constant 200 feet. Adjacent to the beginning of the runway we started another left turn, slowly descending and leveling out just a short distance before passing over the end of the runway. For both takeoffs and landings voice exchanges with the tower were similar to what would take place when operating from a carrier. There was a duty officer at the beginning of the runway who would fire a flare if he didn't like our approach, and we would have to get back in the pattern and go around again.

The only flight I repeated was the check flight before being cleared to solo. I don't recall what the check pilot didn't like, but I didn't have to repeat a flight with Wistle, the usual procedure if the check pilot found a problem. Perhaps we had to come back to the field without completing the check for mechanical reasons, as it was a very short flight and I logged only 1.3 hours. The only difficulty I remember was making dead-stick landings. This maneuver began at an altitude of several thousand feet a distance away from the field. The instructor would pull back the throttle to idle, and with the propellor wind milling I had to judge distance and decreasing altitude to sideslip the plane into the field for the landing. This particular test was usually done at Pace Field, not too far from Whiting. It was a one square mile grass field set among the tall pine trees of the Florida Panhandle. Its shape made it usable regardless of wind direction. Dead-stick landing was one of the tests I had to pass before being cleared to solo, so my check pilot, Lt. Jones, might not have liked my technique. My log book shows a crossed-out solo flight the next day, and then a recheck with another pilot the following day, which I passed. The repeat flight is a mystery that my memory is unable to solve.

Nonetheless, with 26.5 hours of flying time under my belt, accumulated between June 2 and July 11, I successfully soloed in a flight lasting one hour on July 14. At last I was master of the air. Someone wasn't always looking over my shoulder and I could enjoy the freedom of flying, dodging clouds, doing aerobatics, knowing that I was flying the plane and not vice versa.

After soloing I continued flying at Whiting, alternating solo and instructional flights. By this time, since I had repeated only one flight and there were just a few bad weather days when flights had to be rescheduled, I caught back up with my pre-flight classmates and even passed a few. For the next six weeks we trained in two new phases, crosswind and precision landings and aerobatics, with Wistle remaining as my instructor. At the end of each phase there was a check flight with another instructor. I passed both without any problems, and at the end of August transferred to Corry Field to begin instrument training and night flying.

Corry Field - Instrument Training and Night Flying

While training at Corry Field we stayed at the main base at Pensacola. Not having a car when I first arrived, I rode with classmates who did. There was more to do during free time, including visiting a bar in the evening not too far from the main gate with an attractive lady playing the piano. Living conditions were the same as before except for the weekly steak night at the main base officers' club which we usually attended - all the steak we could eat for one inexpensive price.

Corry Field, where we learned the intricacies of instrument flying, had a large number of Link trainers. These were small closed boxes slightly resembling an airplane. The cockpit was configured with the same instrument panel as the SNJ, plus a throttle, stick and rudder pedals. The simulators were free to move slightly left-right and nose up and down, so there was a small sensation that we were moving when we went through the training exercises. Each trainer was connected electrically to a plotting board on which pens recorded the readings of altitude, airspeed, rate of climb, bank, and flight direction.

Once we were seated inside, the instructor closed the sliding cover and we were flying blind with no outside visual cues or references. The only way to survive and not crash was to depend on our instruments. We practiced various types of exercises and then sat with the instructor and viewed the results with the completed exercise shown on the plotting board. The pens never lied! Occasionally there would be a malfunction of the system and we started over again. If the exercise required calling an Air Traffic Control Center or airfield tower for direction, the instructor was in voice communication and performed those functions.

Besides showing our ability to perform different maneuvers relying solely on what our instruments indicated, we trained to navigate over long distances using low frequency (LF) radio beacons. The beacons transmitted a Morse code signal that we heard in our headsets, either "A" (dit dah) or "N" (dah dit), that divided the "sky" into four quadrants. Then, by interpreting the signal, we could decide where we were relative to the beacon. If we were flying away from the beacon, the signal volume decreased. Flying toward the beacon, the signal volume increased. Low frequency beacons were the technology of the day, used to locate our position when flying in bad weather. Each beacon, and there were hundreds located all over the US, had its own unique Morse code identification in addition to transmitting "A"s and "N"s. LF beacons were later augmented by higher frequency beacons that reduced interference caused by thunderstorms and other types of transmissions that made the LF signals difficult to hear when we needed them most.

We alternated Link training with flying. For these flights we sat in the rear seat of the SNJ with a canvas hood pulled over the rear cockpit and went about the lesson of the day. This phase of our training required 24 flights, including two check flights. The instructor sat in the front seat watching his instruments, grading our performance and watching out for other aircraft while we flew the plane from the back seat. A few of the 24 syllabus flights were solo to maintain our landing and takeoff skills. Total flight time in this phase, including a short cross-country flight, was 55.3 hours. Total Link time was 22.5 hours.

The next phase was night flying; now we understood the Navy's wisdom in scheduling instrument flying first. On my first night flight we took off just before dark and circled the field for awhile following the wing-lights and red collision avoidance light of the guy ahead. If we got too far away from the field in the dark, we might not be able to find our way back. Or, perhaps, the Navy was thinking of the poor civilians down below by keeping us away from populated areas in case we crashed. There were a few crashes while I was in basic training; one at Whiting killed the student pilot. When it was dark we made six touch and go landings. They were no big deal but it gave me a real sense of accomplishment landing on an illuminated runway. Night landings required a new adjustment as to when to flare for the landing, using the runway lights to judge altitude. Night flying was not a long phase, just six flights mixed with 14 day flights.

While at Corry, having saved some of my pay, I bought a 1951 Ford Crown Victoria. It was a sporty version of the standard Ford, a two-door model with cream top, a body of robin's egg blue, or Columbia blue (I liked that color), with overdrive. I don't recall how many miles it had logged before I bought it, but I never had any problems over the next two years as I put on a lot of miles. Now, having wheels, I met an attractive redheaded young lady, the daughter of a Navy pilot who was killed during the war. She lived with her mother in a big house close to downtown. I was one of several suitors, but soon after meeting her I moved to Saufley Field so it was a very brief flirtation.

Saufley Field - Formation Flying and Gunnery Training

At the end of November I moved from Corry Field to Saufley Field for formation flying and gunnery practice. It was farther from the main base than Corry, but we were housed on base in similar officer quarters. The Navy built a lot of wooden barracks during WW II, and they weren't going to waste.

First we practiced formation flying. Until now the closest we had flown to another plane was in the landing pattern, and the plane ahead was about a mile away. Now we were trained to fly just a few feet from another plane, usually in a three or four-plane formation. This was really fun as we would rendezvous with the planes in our formation by flying inside a wide circle flown by the lead plane, allowing us to cut it off and quickly catch up. The difficult part was judging when to pull back the throttle and slow, as we were closing at a relatively fast rate. When we matched speed with that of the formation, we slid under it and took a position on the leader's wing.

If you have watched the Blue Angels, they perform this simple maneuver to perfection as well as others we never tried. As we improved our technique the formation leader, an instructor, would lead us through various easy turns and switching positions in the formation using hand and arm signals to indicate what he would do or wanted us to do next. We were required to fly with our cockpit open so it would be easier to bail out in case of a midair collision. Not a bad idea, although I recall only a few close calls. It was November. At 5,000 feet, sitting in an open cockpit, the low temperature made us concentrate on the job at hand.

After receiving an "up check" for this phase we began gunnery practice, always starting our runs from a three-plane formation. We rendezvoused out over the Gulf in a training range reserved for gunnery practice and met a tow plane that was streaming a long banner a few hundred yards astern. From different positions we peeled off and made a run at the banner, firing our two 30-caliber machine guns, one in each wing. We had a simple gun sight mounted on the fuselage outside the windscreen. Before each run the instructor flying the tow plane must have said a short prayer that we would hit only the banner. He also graded our runs, and when we landed the banner was dropped near the runway and recovered to see if it recorded any hits. The shells in each of our planes had different colored dyes so we could be scored when they went through the banner. I have no recollection as to how often, if ever, I hit the banner. But in any case I passed this phase and now considered myself just one step away from being a real Navy pilot, carrier landings. Altogether, 18 flights and 23.3 hours of flight time were added to my log book.

We finished gunnery training near the end of-November and soon after we were granted leave that extended over Christmas 1952. With a former Columbia classmate, Bob Silver, who was just beginning flight training, I drove my "new" car non-stop back to New Jersey. On my first leave since I started flight training I had lots of stories to tell my old buddies. After they graduated from high school, Bobby Sticco and Jimmy Frosco both enlisted in the Navy. They were also home on leave and had their own stories to tell.

Barin Field Carrier Landing Practice - First Carrier Landings

Returning from leave I reported to Barin Field, farther west in the Panhandle, to begin training for carrier landings. Field Carrier Landing Practice (FCLP) was conducted on a small field near Barin. The runway was marked like a carrier deck with a station for the Landing Signal Officer (LSO) at what would be the same place on the fantail of a carrier. Voice calls in the pattern and signals used by the LSO were the same as if we were landing on a carrier. For those who may have seen movies of carrier landings, the mirror system in use today hadn't been invented yet. Instead, the LSO was our mirror system and we depended on him completely to get us down safely.

The procedure for landing on a carrier or during practice landings at an airfield is very simple, the same one we had first used at Whiting Field minus the LSO. The critical part started when we were 90 degrees from the end of the runway and turning toward the LSO. At that point we picked up the LSO, who conveyed his signals with his arms while holding brightly colored cloth paddles (LSOs were nicknamed "paddles"). His signals told us if we were high, low, right or left of a good approach angle, too slow or too fast.

LSO signals are simple to follow. If the approach was on the money, he kept both arms in a horizontal position. If all looked good until a short distance from the end of the runway (in the groove), he gave a cut sign by quickly crossing his right arm over his chest. Then it was up to us to pull the throttle back, stall the plane and flare to a landing. If the LSO didn't like the approach when we got to the groove, he gave a wave-off by rapidly waving his arms over his head. We added full throttle, made a slight left turn to clear the field, retracted our wheels and flaps, and went around again.

We had flown this same pattern from our first flight at Whiting. What was new was watching and responding to the LSO's signals and the stall and flare to the landing. From January 5 to 15 we practiced every day, sometimes twice a day, except for one weekend off. In twelve flights, with a total flight time of 10.6 flying hours, I made 60 FCLP touch and goes; one day I made 20. A touch and go landing means touching down and then immediately adding power and taking off again to rejoin the pattern.

After my 60th FCLP the LSO instructor said I was ready for carrier landings. I drove back to the main base where the USS Monterey CVL-26 was at the dock. Six guys were selected to fly aboard, following an instructor to the carrier. The rest of us were walk-ons who sailed with the Monterey and would switch with the first six and the instructor when they made their landings. The Monterey didn't have to steam too far offshore, and in a few hours the six students and the instructor showed up. The instructor made the first landing, taxied ahead of the barrier, and jumped out with the engine still running.

As I was next in line, I ran out from the island, hopped in his plane, revved the engine to full power per the direction of the launch officer circling his hand over his head, and made a free deck takeoff - pretty exciting for my first carrier takeoff! I was now flying the instructor's plane, which by chance was the same one I had flown for my last eight FCLPs. The instructor had it perfectly trimmed for the airspeed I needed in the landing pattern. The plane almost flew itself, and I made six landings without a wave-off - a piece of cake! After my sixth landing I hopped out and someone else jumped in for his six. Now I was a naval aviator! As a reward for successfully making six landings we were given free drink chits to be cashed at the officers' club bar. I never cashed mine and still have it as a souvenir of my first carrier landings.

For those interested in a little Navy history, the Monterey's designation as a CVL meant that she was a "light carrier," to differentiate her from the larger Essex class. Needing more carriers as WW II Pacific battles increased, the Navy began converting ships being built as heavy cruisers to aircraft carriers. The Monterey's hull was laid down originally for the heavy cruiser Dayton. This hull design was well suited for conversion, for when the Monterey was finally commissioned as a carrier she had a top speed of 32 knots, allowing her to participate with the fast task forces being assembled. She had a length of 622' and beam of 109' at the flight deck, large enough to carry squadrons of the latest Navy aircraft that entered service at the midpoint of WW II such as the F6F Hellcat and the TBM Avenger. The Monterey sailed with the famous Task Force 58 commanded by Rear Admiral Marc Mitscher, participating in many Pacific battles and receiving eleven battle stars. Decommissioned after the war, she was recommissioned in 1950 and stationed at Pensacola for the next five years. If the old girl could have talked before she was scrapped, I expect she would have told many stories of strange attempted landings by student pilots.

Advanced Training

Kingsville, Texas

After completing my carrier qualification I waited for my next assignment to advanced training at Kingsville, Texas. I asked for fighter training, fully expecting to be accepted. But the best laid plans are not always rewarded. I was selected for Anti-Submarine Warfare (ASW) training. I was really ticked and talked to the assignment officer, threatening to quit and go back to the fleet. It was all to no avail. What really bothered me was that I knew I had better grades in basic training than some who were assigned to fighters. I didn't carry out my threat and with a heavy heart reported to Kingsville in mid-February. Some of my original pre-flight classmates got fighter training. My best friend, Frank Garrard, was similarly disappointed and joined me for ASW training. As I recall, the former blimp pilots, those who made it through basic training, all chose multiengine and went to NAS Hutchinson, Kansas.

Naval Auxiliary Air Station, Kingsville, Texas, is in the middle of nowhere about 40 miles southwest of Corpus Christi. The airfield was surrounded on three sides by the enormous King Ranch and thousands of head of cattle. Advanced fighter training was conducted with the F8F Bearcat, a plane I had wanted to fly since I saw one on my first midshipman cruise. ASW training was in the TBM Avenger, at the time the Navy's largest, heaviest single-engine carrier plane. Every time I walked out to my TBM, passing the F8Fs lined up, I was angry all over again, enviously watching my friends getting into the Bearcats. But, resigned to my fate, I learned to fly the TBM Avenger, a WW II torpedo bomber.

If I was to become an ASW pilot why, you might ask, was I learning to fly a monstrous WW II torpedo bomber? It wasn't obvious to me either, although I knew that at the end of WW II convoys were accompanied by small aircraft carriers whose aircraft would patrol and attack submarines threatening the merchant ships.

At our first ground school classes we learned that the Navy had improved WW II anti-submarine tactics and that ASW had become much more sophisticated. This was required as submarines also had improved technology. The most important innovation, the German snorkel, allowed submarines to draw in fresh air for the diesel engines and remain submerged without using their batteries. Previously, submarines had spent a large part of their time on the surface running on their diesels but aircraft patrols stopped that. Nuclear powered submarines were still technology of the future. By using the snorkel they could stay submerged for long periods, making detection much more difficult. And the snorkel, a few feet in circumference, added to the difficulty of detection because it was not easily visible and had a very small radar cross-section.

The Navy's response to this threat was to use a two-plane hunter-killer team. The hunter carried powerful radar that could detect small objects from a great distance. It would then direct the killer to the radar contact. The killer carried depth charges and rockets to attack the submarine. It also carried sonobuoys to drop at the contact point if the submarine had submerged or retracted the snorkel before the killer airplane arrived. With a little luck and crew skill the sonobuoys would track the sound of the submarine running underwater on its batteries and the killer still had an opportunity to drop its depth charges or a homing torpedo. All this equipment was now deployed on hunter-killer TBMs. Also in the works was a magnetic anomaly detection sensor (MAD) that the killer would deploy to better localize the submerged submarine. It worked by recording how the submarine's steel hull distorted the earth's magnetic field. A still larger airplane, the AF Guardian, was under development to carry the MAD and additional sensors.

The TBMs we flew in advanced training did not carry any of the above equipment. Our job was first to learn to fly the airplane and then practice carrier landings and takeoffs. The latter would be done after we finished training at Kingsville. The syllabus consisted of 41 flights that included five at night. In addition we had eight SNJ flights. The first SNJ flight familiarized us with the local area, flight and landing patterns. The next seven were backseat flights to demonstrate that we were maintaining instrument proficiency.

Learning to fly the TBM took some skill. It wasn't a Bearcat, but it was a bear to fly compared to the SNJ. Here are a few statistics on what I would be flying for the next four months. The TBM entered service in 1942 to replace the TBD Devastator, the Navy's primary torpedo bomber until that time. It carried a crew of three, the pilot, rear turret gunner, and a radioman-bombardier who could also man a machine gun under the fuselage in case of a fighter attack from below. It was powered by a Wright R-2600 radial engine delivering 1,900 hp; its top speed was advertised as 276 mph. It could carry the Navy's new 2,000-pound torpedo in its large bomb bay.

It was a very large airplane. It had a wingspan of 54 feet and an empty weight of over five tons. To get to the cockpit, almost 15 feet above the ground, I climbed up using foot and hand holds to reach the wing. From the wing, using one more foothold, I clambered awkwardly into the cockpit, one leg at a time, to sit down on the parachute. The plane captain helped buckle my chute, plugged in my fabric helmet holding earphones and mike (hard helmets were still in the future), and I was almost ready to go.

After starting the engine I spread the wings and confirmed they were locked in place. Because the TBM had large ailerons and elevators in the wings and tail and a rudder almost as big as a DC-3, there was a lever in the cockpit that locked these control surfaces when parked to avoid damage in wind gusts. Going down the preflight check list I unlocked the control surfaces, wiggled the stick, and moved the rudder pedals to show the plane captain the surfaces were free. With his thumbs-up, I was ready to taxi.

My first flight in the TBM was memorable. After takeoff the pattern required a left turn when reaching the field perimeter. Airborne, wheels and flaps up, full throttle, here comes the perimeter fence, roll left. Uh-oh, I couldn't move the stick. Controls locked? No. Ah, I really had to push the stick; this was not an SNJ. Now, with both hands on the stick, I rolled left a little late. The rest of the flight was routine, flying out to the west and becoming familiar with the area. But the lesson was learned; I had to muscle the stick and be prepared. After 1.7 hours in the air I returned and landed. At Kingsville I added another 93.5 hours to my log book.

Carrier Landings in TBM

After completing the above phase of advanced training that included a cross-country flight to Tulsa, we were shipped back to Barin Field in mid-May to start FCLP practice. Being back at Barin Field was "old hat," but now I was flying a much bigger aircraft that required better flying skills. However, as we were already carrier qualified, the Navy didn't expect we would need a lot of training to go back onboard the carrier. We practiced for six days; with 62 FCLPs and 15 more flight hours I was declared ready. This time I flew out to the Monterey, made eight landings, seven free deck takeoffs, one catapult launch, and flew back to Barin Field. No problem! I received another free drink chit that I didn't cash. I wonder if they are still good. On June 3 I was given Naval Aviator Certificate number T-3351 and the wings of gold were pinned to my chest by NAS Pensacola's Commanding Officer, Captain G. S. James, Jr..

Advanced Instrument Training - Corpus Christi, Texas

Now we were on the last lap before reporting to an active duty squadron. In July we went back to Texas, this time to NAS Corpus Christi All Weather Flight School. It would be a real change of pace as we would be trained in the small twin-engine Beechcraft SNB Kansan. In addition we would be entombed again for many hours in Link trainers but for much more intensive training than what we had received at Corry

Field. If we passed all the requirements we would receive a Standard Instrument Rating that would allow us to fly the commercial airways in instrument conditions when we got to our active duty squadron.

The flight syllabus consisted of 21 flights that included four at night. For each flight three students joined the instructor and alternated taking the copilot seat, flying under the hood for instrument training. It wasn't very exciting, as the instructor made both the takeoff and landing, but I wasn't too interested in flying this airplane. The real work was in the Link trainer; by the end of one month I had completed 180 hours sitting in the little boxes and qualified for an instrument rating. Upon completion I took three weeks of leave, returning to my parents' home in New Jersey. While there I bought the aviation green uniform worn only by pilots, aircrew officers and chiefs.

From the time I started gunnery training Frank Garrard, who was in the same pre-flight class back in April 1952, and I had been joined at the hip (and shoulder) as we progressed through the various phases. He had been ahead for a short time because of my ear problem, but I had caught up. We roomed together at Kingsville, carrier qualified in the TBM at the same time, and received our wings the same day. Now we had finished All Weather Flight School together, often flying with the same instructor, and were assigned to the same active duty squadron, VS-30, at NAS Norfolk.

Frank was a Georgia Tech NROTC graduate whose home was in Columbus, Georgia. Short and stocky, with a pronounced southern drawl, he had been on the Georgia Tech wrestling team. On one long weekend during training at Pensacola we drove to his home, a large, white colonial-style house. When we arrived we were met at the door by a plump, older black woman who gave him a big hug and with a smile followed us into the house. Frank told me that she had been the family cook as long as he could remember. In 1952 some parts of Georgia had not changed much from Civil War times when, based on what I had read, black household staff were often considered members of the family.

SNB

SNJ

TBM

Dilbert Dunker

On leave with my 1951 Ford (the original) at Bobby's Sticco's home

Intrepid airman Beattie at Barin Field, Florida, December 1952

Advanced flight training class, Kingsville, TX. On TBM wing, back row, Frank Garrard, Earle Callahan, Rex Whitcher, Bob Kowalski, tall guy, Jim Winnefeld. Instructors Lts Harry Johns and Burt Lewis in front, May 1953

Pensacola CO Captain G. S. James Jr. pinning on my wings, June 1953

Chapter V
Fleet Squadron VS-30 - 1953 to 1956

Reporting to VS-30

Frank and I reported to VS-30 at NAS Norfolk, Virginia, commanded by CDR John David Howell, on August 18. We roomed together again in the BOQ. We learned we had joined a recently reactivated reserve squadron, VS-801, that included a number of recalled pilots, some in their forties and not too happy. In fact my first eight flights, logged at the end of August, were recorded as VS-801 training flights. The good news was that the squadron had just transitioned from the TBM to the AF Guardian, so my first flights were in this new aircraft. One of the several VS squadrons at Norfolk was still flying versions of TBM hunter-killer planes.

The TBM was a large airplane; the AF was even larger. It was a typical product of the Grumman "Iron Works." Grumman airplanes were known for their ruggedness and ability to withstand damage. The AF was no exception, and it supplanted the TBM as the Navy's largest single-engine carrier aircraft. It had a wingspan of a little over 60 feet, a length of 43 feet, and empty weight of just over seven tons, two tons more than the TBM. It was powered by a Pratt & Whitney R-2800 radial engine delivering 2,400 hp with an advertised top speed of 315 mph (perhaps achievable in a steep dive). It was another bear to fly, although it was easier to control than the TBM as it had hydraulic boost on the control surfaces.

Aerobatics were not allowed according to the flight manual; but who paid attention to all the do's and don't's when he was only 24? One day I tried making a slow roll, an easy maneuver. As I rolled over on my back at full throttle, I quickly decided it was a bad idea. Months worth of debris that had been accumulating on the deck and below the seat, or caught in the instrument and control panels on either side of the cockpit (for some long flights we carried and ate from lunch boxes), came raining down on my head. I completed the roll in a steep swooping dive and never attempted it again. I never tried to do a loop. I figured at the top of the loop the plane would stall out and I would end up in an inverted spin, definitely not a good position to be in. Through September I made 24 flights in all three versions of the AF, the hunter with the big radome in the belly and the two killer versions, with and without MAD gear.

Carrier Landings in the AF Guardian

At the end of September and into early October all the new pilots began preparations for carrier landings. We practiced at Fentress Field in the pinewoods south of NAS Oceana, Virginia. After 78 FCLPs our senior LSO, LT Bill Hackett, thought I was ready to land on the escort carrier USS Mindoro CVE-120 that would be cruising off the Virginia Capes. She was smaller than the Monterey where I last landed; the landing area between the fantail and the barrier was about 50 feet less.

It was tight, but we had been practicing at Fentress with this in mind. Probably this is one of the reasons why I made 16 more FCLPs to practice landing the AF on the Mindoro compared to the number of FCLPs I needed to practice before landing the TBM on the Monterey. I should add that our two LSOs, both squadron pilots themselves, were exceptional, giving us great confidence in their signals during landings. I made my first AF landings from October 6 to 15, without any problems.

My log book indicates I was considered day qualified on October 15. I had made 20 carrier landings and five catapult shots. Our catapult launches were really shots. The steam catapult now in use hadn't been invented. On the Mindoro the catapult energy used to get us flying was provided by a hydraulic ram. The good news was that the Guardian flew comfortably at 85 knots, well within the ram's ability with the engine running at full throttle to accelerate the AF from zero to 85 knots in less than 200 feet by the end of the deck. With luck there might be a strong headwind to provide additional airspeed margin.

I would make several deployments on the Mindoro. The designation CVE indicated she was an escort carrier, the smallest and slowest in the fleet. The Navy needed more and more carriers in WW II thus many were built in this smaller class, some originally laid down as other types of ships including tankers. She was commissioned in December 1945 so saw no action in WW II. Her size, a little over 557 feet in length, meant she was smaller than the Monterey and much slower; at full speed she could make only 19 knots. Her flight deck was 501 feet by 80 feet so the AFs 60-foot wingspan made for tight takeoffs and landings.

A carrier's speed is important in many ways, but from a Navy pilot's perspective it is very important. If the prevailing wind during operations was low, the Mindoro steaming into the wind couldn't significantly increase wind speed coming across the deck. The higher the wind speed, the greater the safety margin for both landings and takeoffs when a plane is flying just above stall speed. Large modern angle-deck carriers that can make 32 knots or more don't have this problem because, among other considerations, if the pilot misses the arresting wires he can add full power and immediately be airborne again. There is no barrier to run into, and for standing takeoffs the steam catapults have more than enough power to quickly accelerate the heaviest planes to flying speed.

We didn't have that luxury. When the LSO gave us the cut signal we stalled out and landed. If we landed long, we hit a steel cable barrier and damaged our propeller and engine. The barrier was always up during landings because planes were parked on the deck forward of it. Today's Navy pilots don't know how good they have it!

Carrier landings involve other considerations in addition to those explained. While you are making an approach the ship is pitching up and down, and in rough seas the deck elevation might change by tens of feet in a very short time. This required the LSO to judge how high or low the deck would be when he signaled cut. Perhaps he would give you a wave-off even though the approach was good to avoid making a landing while the deck was falling away or rapidly rising. Depending on the sea state the carrier would develop a rhythm in her pitch, and between the ups and downs there would be a period when the pitching smoothed out. If the timing of the approach was good relative to a level flight deck, the LSO would be more likely to give a cut. Landing during a heavy sea state could be difficult, even more so on an escort carrier whose length is about 300 feet shorter than that of the Essex class. The longer the carrier the less it pitches and the longer the time when the deck would be relatively level.

ASW Training and New Flying Skills

Back at Norfolk, for the next two and a half months night and day training flights were frequent and included ASW exercises. Also I was introduced to a new instrument landing procedure, ground-controlled approach (GCA), the Navy equivalent of an instrument landing. A GCA ground controller would pick me up on his radar some distance from the field. Using voice commands he would guide me to a landing with constant updates on altitude and speed, and instructions on when and which compass heading to take. I just followed his directions. His last communication would be, "Runway should be in sight," and I would make a normal visual landing. If at that point I thought I wasn't in good shape to make a landing, I would wave off and start over again. This was the same procedure that would be used at sea with the GCA controller in the ship's Combat Information Center. In bad weather, when a pilot couldn't see the carrier until just a few hundred yards away, the controller would direct him close astern and into the proper position to pick up the LSO. Then the LSO would give the normal signals to the cut. Nothing to it - it just took a little practice.

Our next flights began rocket firing practice from killer planes at a ground target located near the beach at Duck, North Carolina, north of Kitty Hawk. The target was a set of concentric circles marked on the ground. A firing run began by reporting in our flight number, e.g. "Scorebook one zero" (Scorebook was our squadron call sign), and start a shallow dive from west of the target toward the ocean. At about 500 feet, with the target in sight, we fired and pulled up, circling back to the west. An observer in a bunker recorded our firing.

Night practice was the same with a few small flaming pots outlining the target. A crewman in the back seat focused our searchlight on it to improve our aiming. This was lots of fun, but one night we had a tragic accident. One of the new pilots, Henry DeJan, failed to pull up after firing and flew into the ground, killing his two crewmen and himself, the only fatal accident in the squadron while I was assigned.

We did have one near-tragic accident. On a cross-country night flight Frank Scott, with two passengers tagging along for the ride, had a fire. They couldn't put it out, and somewhere over the West Virginia mountains they all bailed out and landed safely. There was a humorous side to this incident. One of the passengers was the squadron's flight surgeon Doc Ingram, onboard to get his monthly hours to receive flight pay. Flying was not his favorite way to pass time, so he seldom flew. He might have had a little picker-upper to calm his nerves, something he was known to do. Anyway, he bailed out and lived to tell the tale to his grandchildren. I'm not sure that he ever flew again, as he soon retired.

Night Carrier Landings

In January 1954 I began night FCLP practice. In two nights I made 20 landings and was ready to go. On the 27th we flew again to the Mindoro, steaming off the Virginia Capes. Over the next few days I made six refresher day landings and catapult shots. On February 1 I made my first night landing and seven more the next four nights, then flew back to Norfolk. The only really exciting event was Joe Bajak's first night landing. After taking the cut he didn't flare early enough to land on the deck and bounced over the barrier. He added full power and became airborne, flying past the ship's bridge and scraping off about ten feet of his starboard wing. The AF kept flying, and he was directed back to Norfolk. He never again tried to become night qualified.

There were a few differences in making a night landing compared to day landings. As we approached the ship flying downwind at night, we initiated our turn in by judging our position relative to the red truck lights of the plane-guard destroyer (truck lights are placed at the highest point on a ship's masts) that was steaming 1,000 yards or so on the carrier's port side. She was there to make an immediate rescue in case of a crash during the approach.

The colored stripes on the LSO's flight suit legs, body and arms that stood out in daylight were made of a material that glowed at night as he stood in front of a UV light. From a distance he looked like a stick figure as he went through his signals. The carrier was completely dark except for its red truck lights and a few tiny white lights down the flight deck center line. When the LSO gave the cut he stepped on a switch, turning on lights along the deck edge that illuminated it and gave us a reference for when to pull back on the stick, flare to make a three-point landing, and catch a wire with our tailhook. As soon as we were down the LSO released the switch and the deck was dark again. Following the directions of the flight deck crewmen using their lighted wands we raised our tailhook and then they directed us to our parking spot. Once clear the barrier was raised for the next landing. Normally that would be the end of the flight, but, as we were practicing landings, we taxied to the catapult, hooked up, and immediately relaunched.

One other major difference required greater pilot skills for night landings. During day landings if the barrier operator saw us catch a late wire with our tailhook he could quickly drop the barrier and let us roll over it, avoiding damaging the plane. At night this wasn't possible as the barrier operator couldn't be sure if we caught a wire and couldn't drop the barrier until he was sure. This meant that at night we had one less wire to catch, making the landing area 50 feet shorter. For this reason night landings required greater proficiency, and in the old "straight-deck" carrier Navy not all pilots became night qualified. I loved night landings because the adrenaline flowed and I could demonstrate my piloting skills for all my shipmates. After completing flight operations, day or night, the LSO came to the ready room and reviewed our landing technique or lack thereof for all to hear.

From February through the end of April we made three more short deployments on the Mindoro that included night searchlight practice. Both killer versions had a two-million candlepower searchlight on their port wing to illuminate surface targets. It was controlled by the crewman in the back seat, who had a joystick to point the searchlight by looking through a periscope mounted underneath the plane. We made more night landings, preparing to sail on the Mindoro and join the Sixth Fleet in the Mediterranean.

Sixth Fleet Deployment on USS Mindoro

On May 13 the Mindoro left Norfolk and steamed across the Atlantic with a small task force. Once underway our days were either boring or exciting, more often boring as we seldom flew two days in a row and often not for a week or more. Our daily routines were rather erratic as we didn't stand a normal four on and eight off duty cycle as did the ship's officers.

My primary duty was as a pilot, but before leaving I was given a new collateral duty as squadron communications officer, relieving the officer who was leaving the squadron. Among other responsibilities I kept all the classified publications up to date and at preflight briefings handed out the codes for that day and instructed the crews in their use. At the end of each flight I collected and secured them in a safe. In anticipation of being relieved my former squadron mate had become delinquent in his duties; I inherited a safe containing many out-of-date publications and spent the first week at sea cleaning up my new empire.

The ensigns and junior-grade lieutenants were berthed together in two large compartments in the fantail under the hanger deck. The more senior officers bunked two to a room. With six of us living together in close confinement with no portholes, at times our quarters became a little gamey. Laundry could be given daily to the stewards, but returning late in the day or at night we would hang our sweaty flight suits near our bunks. Depending on our schedules we might wear them several times before they saw the inside of a washer.

Probably all pilots are somewhat superstitious or have little idiosyncrasies that comfort them as they take to the sky; we had one pilot, Jim Winther, whose superstition affected all of us in our home away from home. For good luck he refused to wash his flight suit, and it hung for several weeks near his bunk. One day while he was eating we tied it to a long rope and threw it overboard. It was dragged astern until he returned. Irate, he found what we had done and rescued it, a little the worse for wear but still usable. But he forgave us and became a little more diligent in his hygienic practices in the weeks ahead.

We spent our free time either in the ready room or wardroom. Reading was my favorite activity, but we also had competitive bridge, acey-deucey, backgammon and cribbage tournaments. My bridge partner Charley Butler and I became very adept, in addition to conventional bidding, in signaling our hands; some purists might have called it cheating. But it was all in fun and without cash prizes. Because our flights often didn't conform to the ship's normal routine, when we returned we were able to go to the wardroom day or night and order a meal, a welcome reward for hazardous duty.

At the beginning of our deployment with the Sixth Fleet we conducted a number of exercises with different units. We flew day and night missions, all starting with a catapult shot because the crowded flight deck didn't permit a free deck launch. Our first liberty was outstanding as we visited Gibraltar and from there took shore leave to Seville, Spain, with a few intermediate stops.

From Gibraltar we sailed to Barcelona, where the squadron basketball team that I coached with Roger Visser was invited to play a local team (Roger was also a player; I wasn't good enough to make the team). The arena was the bullring in which a court was outlined in chalk powder and two baskets installed. We beat the Barcelona team and were taken to lunch at a local restaurant. It was a welcome change of pace to act as friendly ambassadors (perhaps we should have lost the game) while the US sought to improve relations with Spain during the Franco regime. During lunch I asked what I thought was an innocent question about a restrictive government tourism policy I had read about; an embarrassing silence followed and one of our hosts quickly changed the subject. Junior officers weren't expected to be wise in the world of diplomacy when dealing with an authoritarian government.

From Barcelona we steamed to Naples and a few of us took a tour to Rome. Bob Kowalski and I met two attractive young ladies and spent most of the day with them sightseeing and trying to communicate using my poor Spanish and their limited understanding of English. We made our final port visit in Genoa and then headed back to Norfolk, all this in just nine weeks. Join the Navy and see the world!

The most memorable event of the cruise occurred on the night of July 13. While steaming west of Sicily, on our way back to the States, we conducted a normal four-plane night patrol with two pairs of hunter-killers sweeping ahead of the task force. Frank Garrard and I were a team, Frank in the killer and I in the hunter. Returning to the Mindoro after about three hours, we all joined up over the ship and took a standard break to come in for landings.

Frank was first in the pattern and I was second. When I made my turn in at the 180 I could see Frank's wing lights up ahead as he turned in for his cut. When I reached the 90-degree point where I would pick up the LSO, the ship should have been dark but the deck lights were still on. Maintaining my approach but still far from the groove, I got a wave-off. Adding power and retracting my wheels, I flew past the port side and caught a glimpse of Frank's plane in the starboard catwalk but couldn't see what the problem was.

Word came for the rest of us to join up and orbit the ship until further notice. Frank had landed too far to the starboard side of the deck; the arresting wire had pulled him farther right until his wheel went into the catwalk and he flipped over. That was the bad news. The good news was that the arresting wire held and his plane was suspended upside down over the starboard side. With the arresting wire still holding Frank's plane further landings were impossible.

What should the three planes still flying do? We were too far away to make a landing field on shore. The Mindoro crew would try to get the deck ready for our landings but there was no assurance that this could be done before we ran out of fuel. We had to be prepared to ditch alongside the plane-guard destroyer.

We orbited for the next two hours, watching our fuel gauges stretching out the fuel remaining and, from a distance, what was happening on the brightly lit deck. Good news again, the Mindoro crew managed to secure Frank's plane and, with one less arresting wire to use, we all came in and landed. The other good news was that no one in Frank's crew was injured; they scrambled back on board while the plane was hanging inverted over the water.

But a problem surfaced after I landed. While Frank's plane was upside down the classified publications he carried to transmit coded messages fell into the water. When returning to the ship at the end of a flight the required procedure was to put the publications in a weighted canvas bag against the possibility that the plane would crash at sea and the publications retrieved by the "bad guys." Frank hadn't done that, so they were presumed to be floating somewhere in the Mediterranean Sea. Our skipper, CDR Bigelow, was very upset. It was known that Soviet submarines shadowed Sixth Fleet task forces to observe operations. Therefore, the classified publications that had dropped out of Frank's plane could now be in the hands of our Cold War enemy.

As the squadron communications officer and in charge of classified publications, it was my responsibility to make sure the pilots observed the rules and stowed the publications in the weighted bags. The problem was my fault. Bigelow confined me to quarters and for the next two days my constant companions were some out-of-date publications floating in a bucket of salt water. The question: how long would it take for them to sink? Of course they never did and one night a couple of "friends" swiped them. Soviet spies were everywhere! This was all in good fun, as we thought the test was comical. I'm not sure Bigelow ever fully forgave me, although he gave me a few outstanding grades in my fitness reports after the incident, so perhaps he did.

Norfolk - New Romance

Back at Norfolk by the middle of July we continued routine training exercises. I bought a 1954 Oldsmobile Super 88 hardtop, cream above and forest green below, the latest in design with wrap-around windshield, automatic transmission and more. It was a beautiful car, worthy of a sharp young Navy pilot. At the same time, without realizing it, I met my future wife. I mentioned that Frank Garrard and I roomed together at the BOQ. Actually we were in a two-bedroom suite with a bathroom in between. In the other bedroom were Bob Kowalski and Jhea McCloskey, two pilots who had reported to VS-30 at the same time as Frank and I. Jhea had a cousin, Patricia McCloskey, who lived in Stamford, Connecticut.

With Jhea providing the introduction, on one of my occasional visits back home to New Jersey we arranged to meet. Thus started a rather long-distance romance. At close of business on Friday afternoon, when I wasn't scheduled as the duty officer on the weekend, I would race over to the Cape Charles ferry (the tunnel wasn't even a vision in 1954), up the Delmarva peninsula on Route 13 to stay with my parents in New Jersey, and then on to Connecticut to visit Pat.

We might go into the City for a Broadway show or dinner and dancing at the Starlight Roof of the Waldorf Astoria. I was really trying to make a good impression based on my vast worldly experience. Expense on the dates was not very important. What else was a bachelor pilot to do with his flight pay? I must have succeeded; as the romance flourished I began staying overnight at Pat's house instead of going back and forth to New Jersey. Then I would race all the way back to Norfolk to make muster on Monday morning. I was young, and a lingering kiss was worth more than a good night's sleep.

In July we learned that within a few months the AFs would be retired and the squadron would begin flying the new Grumman S2F Tracker. Flying a twin-engine airplane off an aircraft carrier day and night, with a junior pilot sitting in the right-hand seat, was not my idea of a Navy flying career. I wouldn't be due for a shore assignment for several years, something else that wasn't very appealing. I requested a transfer to the Naval Air Test Center Test Pilot School at Patuxent River, Maryland. At the same time, and for the same reasons, Frank Garrard requested a transfer to a fighter squadron. Frank's request was approved; mine wasn't. Considering my alternatives, I decided to change my status from regular Navy (USN) to reserve (USNR) and start planning a life outside the Navy when my required active duty was completed.

Operation Blackjack

With only a few more months remaining to fly the AF, at the end of

July the squadron received an unusual request. A sister squadron stationed at Quonset Point, Rhode Island, VS-39, was preparing to deploy on a NATO exercise in the North Atlantic. It was short of night qualified pilots and requested four volunteers. This sounded pretty exciting. Volunteers didn't have to complete the VS-39 deployment on the USS Valley Forge CVS-45. The squadron would stay onboard after the NATO exercise and join the Sixth Fleet in the Mediterranean. If desired the volunteers could be released at Belfast, Ireland, after the NATO exercise and proceed back to Norfolk independently. I convinced three other squadron mates, Bob Kowalski, Bob Timm and "Speedy" McCall, to join me. Another attraction was that I would be the senior officer in charge of the detail (the other three were ensigns) and would decide when we would schedule the "independent" return. Visions of liberty in Ireland and England "danced in our heads."

At the beginning of September, carrying all our gear, we flew by military airlift to NAS Quonset Point to meet our new squadron mates and participate in the annual NATO exercise, Operation Blackjack. The Valley Forge was an Essex class carrier, the largest in service during WW II, larger than any we had flown from up to that time. We sailed from Quonset Point on September 7 and were tested immediately on our ability to make night carrier landings as the Valley Forge prepared to join the rest of the ships gathering near Ireland. Over the next week, according to my log book, I made three day and three night landings. I assume the other three did the same. Bob Kowalski reminded me that, on the first night mission we all flew together, we landed at 20-second intervals without incident or wave-offs, not bad considering we were flying off a straight deck carrier. Every time a plane landed the barrier had to be lowered for the plane to taxi over it and then raised again before the next plane was given the cut, flight deck teamwork at its best. By the time the four of us went below, the Admiral posted a message from the bridge on the large briefing screen at the front of the ready room congratulating us on our excellent piloting. Of course, what else could he have written?

We rendezvoused north of Ireland with the rest of the NATO participants and steamed toward the Faeroe Islands, conducting continuous anti-submarine exercises. This was the area of the North Atlantic transited by Soviet submarines after they left their base in Murmansk, monitored all the while by the SOSUS acoustic network sensors on the ocean floor. We flew almost every day or night for the next two weeks, and during one of the exercises I was credited with a kill when I dropped a Mark 15 homing torpedo on a friendly submarine, without its warhead of course.

All the flights were tedious. Operating in the North Atlantic in September and October we were required to wear anti-exposure suits as water temperatures were below 50 degrees. The outer layer of the suit was nylon-coated rubber with attached heavy rubber boots. It was similar in appearance to a deep sea diver's suit minus the helmet. The inner separate layer was a heavy quilted one-piece zippered garment that covered us from the neck down. After donning the quilted suit we climbed into the outer layer, stepped through a slit in the front, pulled up the bottom half, then pulled the top half over our heads. This half made a tight seal around neck and wrists. The slit was closed by rolling up the rubber bladder and sealing it shut. On top of the suit we wore a standard inflatable life jacket with attached dye packs, shark repellent and small light. Altogether they weighed about 40 pounds. By the time we got in the cockpit after a one-hour preflight briefing we were sweating. Once we were in the suit sweat could not evaporate. If we ended up in the drink it was supposed to keep us alive for about one hour, depending on water temperature. Fortunately none of us ever needed to test the suits' effectiveness.

Only two incidents of note occurred during Blackjack, both at night. Soon after one launch I had an electrical fire. I popped all my circuit breakers except those for my instrument lights and radio and reported the problem. I was ordered to return for a landing. I explained what I had done and, since there was no longer any smoke in the cockpit or crew compartment, suggested that I just orbit the task force and land with the returning flight. The Admiral must have liked that idea, for to bring me back right away would have required reorienting the destroyer screen and other ships so that the Valley Forge would be steaming back into the wind. I was given permission to orbit overhead and for the next three hours flew a racetrack pattern over the task force, landing when the other three planes returned.

The other incident, somewhat amusing, involved a short exchange with my wingman Bob Timm. One night while flying a normal search pattern far ahead of the carrier he called in a subdued voice, "Don, there is a UFO flying off your starboard wing. Do you see it?" Of course I looked quickly, never having seen a UFO, but I saw only my own wing light. I reported this to Bob; he didn't answer and the rest of the flight continued observing standard radio silence. Every now and then at squadron reunions I remind Bob of our close encounter of a friendly type.

Blackjack completed, the Valley Forge returned to Belfast. We four had agreed that, as VS-39 had no further need for additional night qualified pilots, we would leave the ship to visit Ireland. The first thing we did after going ashore was buy some clothes that would pass as civvies.

I had a difficult time finding a sport jacket that fit, and the one I finally bought was a little short in the sleeves. I guess in 1954 there were not too many men over six feet in Northern Ireland. A bigger problem was shoes. I wear a size thirteen and couldn't find any that size, so I wore my plain-toe black Navy issue, not very elegant but probably no one noticed.

After our shopping spree we set out to see the sights. We met some Irish lassies who helped introduce us to the countryside. I remembered my father saying that the Beatties had come to the US from an Irish town named Bailieborough. I bought a map and, sure enough, there was a town by that name just south of the Northern Ireland border. I talked Bob Kowalski into joining me, and with two young ladies in tow I rented a car and we drove south.

Bailieborough turned out to be a very small town, just a few houses and a church on a hill. I drove to the church, figuring that the parish priest might be the best source of information on any Beatties still living there. He was most accommodating; he said that there were two spinster sisters in the parish named Beattie and pointed out their house. I parked in front and knocked on the door. A small elderly lady answered. After a brief introduction I explained my quest. "No," she said, "I don't know of any relatives that emigrated to the US." As she wasn't especially friendly, I thanked her and we drove back to Belfast. If my father was correct on the location of his great grandfather's home, perhaps the passage of over 100 years had erased any memory of his departure.

After a few more days in Belfast we caught a military flight to England and headed for the bright lights of London. We were able to get rooms downtown at the Air Force Columbia Club BOQ. From there we visited Buckingham Palace to see the changing of the guard, the crown jewels and more, and ate at some great restaurants. We didn't try too hard to return to the States, but after two weeks we ran out of money and caught a military flight back.

Transitioning to the S2F Tracker

Back at Norfolk we found some new S2Fs parked on the tarmac. I made my last three AF flights on October 28 and one cross-country flight on November 2 and 3. On October 29 I was checked out in the S2F and made four more flights by the end of November, including one night flight; goodbye Grumman Guardian, hello Grumman Tracker.

The S2F, a twin-engine carrier plane, combined in one aircraft both the hunter and killer functions that required two AFs. It was a rather large plane with a wingspan of 69', a length of 42' and a height of 16'.

The engines were Wright R-1820s, each delivering 1525 hp, giving it a top speed of 280 mph and a range of 1,350 miles. Empty weight was a little over nine tons, so it was heavier than the AF. It carried a crew of four, two pilots and two crewmen who operated the retractable radar and MAD gear and monitored the sonobuoys when they were dropped. It had a small bomb bay that could hold two homing torpedoes. Racks on the wings could carry rockets and depth charges and a wing-mounted searchlight that was operated by the copilot. Altogether it was a formidable platform that efficiently carried out ASW operations that formerly required two-plane teams. But in my mind it was still a dog.

Squadron conversion went smoothly, and in November and December I made 26 day and night flights, reworking all the ASW procedures I had learned in AF hunter-killer teams. But soon after the S2F's introduction in the fleet there was a major problem. During takeoff the wings on an aircraft undergoing tests at Patuxent River suddenly folded. I don't recall the result of the accident, but it required a change in procedures we followed when spreading our wings prior to taxiing (while parked, wings were almost always folded to conserve space). After spreading the wings, the locking system at the joints required 16 short pins in each wing to set and lock the folded wing to the wing stub attached to the fuselage. After the accident a small hole was cut in each wing above the locking mechanism so that crewmen could look inside and confirm that the pins were in place. As both crewmen had overhead escape hatches in case of ditching, to complete the required inspection they stood on their seats and looked in the hole where the pins were visible. This wasn't hard to do, and after that change I don't recall any further problems.

Flying the AD Skyraider

To assuage our displeasure at flying another low and slow airplane, and with a junior pilot in the right-hand seat adding to our annoyance, Bob Kowalski and I came up with a happy solution. One of the sister squadrons at Norfolk was a Fleet Air Support Squadron (FASRON). Its major purpose was to allow pilots assigned to desk jobs to maintain their skills and, by so doing, keep receiving flight pay. For this they had to log six hours of flight time a month. As these desk-bound pilots came from many different squadrons, their recent flying experience had been in many types of airplanes. The FASRON had jets, props and a few multiengine planes.

In the inventory were several AD Skyraiders, a fabulous, high powered, versatile airplane with a seat for only one pilot. We checked out in the AD without telling our skipper and sneaked away for our flights. It was a simple checkout consisting of studying the manual, answering some written questions, and going through a blind cockpit check in which the check pilot asked us to place our hands on different levers or buttons. This demonstrated that, if a problem restricted our vision or we lost cockpit lights at night, we could put our hands on critical items like the landing gear or flap levers to make a landing.

In return for being allowed to fly the AD we cooperated with the local CIC and GCA schools that were training students in controlling airplanes. At the beginning of each flight we reported in to the schools, and for the first hour or so were vectored all over the local area responding to the commands we received from those in training. It was pretty simple; we just had to be aware of other traffic in case the trainee missed observing another plane on his radar.

The AD was a high-powered plane with a Wright R-3350 compound engine with two banks of nine cylinders delivering 2,700 hp, giving it a maximum speed of over 300 mph. The FASRON planes were stripped-down models, much lighter than those flown in the fleet, so they could really go. They were a little tricky to fly, and if we weren't careful as we taxied to the runway some of the spark plugs might foul and we couldn't get a good carburetor check. To avoid this we taxied with the mixture in lean and put it in rich only when we were ready to go through pre-flight checks. But, once airborne, the plane was a gem.

After completing the school training period we were free to fly around for the next few hours. I would fly over to NAS Oceana, the Navy master jet field, and sometimes entice a jet pilot into a dogfight. Great fun! If he didn't get a good shot on his first pass, I had the advantage; I could always turn inside him with enough speed to make a firing run. His only alternative was to add power and, with his superior speed, get away. This was all simulated, of course. However, during the Korean war ADs shot down Soviet MIGs, demonstrating their ability to contest the sky with a jet. During one flight, after being released from GCA training, I flew up to Stamford, identified Pat's house, and did some slow rolls. As I recall her family saw me and I thought it was worth the risk to fly so far out of the Norfolk area.

Having made the decision to leave the Navy as soon as my time was up at the end of 1955, in mid-December I flew to Floyd Bennet Field and took a cab to Columbia to talk to some of my old profs. I asked their opinion on returning to Columbia to earn an advanced degree in geology. I told them I wanted to do field exploration either in the US or abroad. Their advice was somewhat surprising. They recommended going to another school for graduate work to broaden my experience, specifically the Colorado School of Mines which I had never heard of. I decided to make a visit as soon as possible.

Back at the squadron, for the next three months we were busy learning to fly the S2F in all types of ASW exercises and conditions, day and night flights, searchlight runs, GCA practice, homing torpedo drops, rocket firing and even FCLPs although we didn't have a carrier deployment on the schedule. I took a few cross-country flights to Lakehurst, Floyd Bennet, New Orleans and Key West. The squadron was required to log cross-country hours. Since the married pilots were reluctant to make these flights on weekends, the bachelors took advantage of the hours available and did a little sightseeing. The S2F was not a difficult plane to fly; it had plenty of power and another pilot in the right-hand seat made cross-country flights easy as he helped keep track of the navigation.

Our new Executive Officer, LCDR Bob Gates, arrived with the S2Fs. Bob's last assignment had been at the Naval Air Test Center, Patuxent River. I was assigned to help him transition to the S2F, and we flew several flights together including his first FCLPs. During one flight I asked him about his time at Patuxent River and mentioned that I had asked for duty there but had been turned down. He said he thought he could get the assignment for me. Bah humbug - too late! I had already begun the process both mentally and officially of leaving the Navy. I have always wondered what my career would have been if I had taken him up on his offer.

We received some sad news about this time. Frank Garrard, during one of his transition flights in an F9F-8 in his new fighter squadron, had crashed and been killed. The cause was attributed to a malfunction of the "flying tail." On the F9F series, to provide more control in high-speed flight, the entire horizontal tail could be moved - not just the elevators. The pilot engaged the "flying tail" by punching in a knob on the right console. It was believed that when Frank punched the knob the "flying tail" malfunctioned and put him in a dive from which he did not recover. Sad as the news was, I knew that Frank at least had had the pleasure of experiencing a few flights in a fighter, his dream as well as mine from the time we first reported for flight training.

At the beginning of April we started preparing for our first carrier deployment in the S2F. We were scheduled to make landings on the USS Antietam CVS-36, the first US carrier to be converted to an angle deck, a British invention. This would be a new experience in more ways than one.

After the usual FCLP practice, eleven landings, we flew out to the Antietam on April 12. I made twelve landings that day, eleven touch and goes and a final one to allow another pilot to switch and try his luck. The next day I made eight more. The S2F was a well-behaved plane during carrier landings and takeoffs, and I had no problems. The angle deck really simplified landings, as we had the benefit of a large deck and no barrier to worry about. If we didn't catch a wire we just added power and took off. In May I qualified in the S2F for night carrier landings.

The only difference I noted during landing on the Antietam was that, when lining up for the cut on the angle deck, I flew through turbulence coming off the ship's superstructure. For landings on straight deck carriers the ship steamed a little off the direction of the wind so that air turbulence from the superstructure was off to starboard as we came straight in. But the angle deck meant overshooting the ship's direction and coming in on an angle on the starboard side before taking the cut. The turbulence wasn't excessive, but for the first few landings it caught me by surprise. I assume that for jet landings this isn't a problem, as they are taking a cut at much higher speed and quickly fly through the turbulence.

Marriage

During April I took another major step; Pat and I became engaged. Our wedding was scheduled for August. But before that momentous occasion I flew a full schedule of squadron operations on the Antietam, including mentoring some new pilots. I also took the FAA tests and qualified for a single and multiengine commercial license, which might come in handy as I hadn't yet committed to returning to graduate school.

On August 14 we were married at the Catholic church in Port Chester, New York, Pat in a beautiful white wedding dress and I in my dress whites. We were accorded full military honors as we left the church, walking under the crossed swords of Jhea McCloskey, Bob Kowalski, Dave Hunt and "Speedy" McCall. After the reception we left in the Oldsmobile for a tour of New England. The second night we stayed with a former squadron mate, Jack Lund, and his family in Marblehead, Massachusetts. From there we went to Bar Harbor, Maine, and circled west to Niagara Falls. After a week we returned to Stamford, loaded the car with Pat's wardrobe and wedding gifts, and drove to Norfolk. I had rented a small furnished house in Ocean View, just a few minutes from the base. Life as a married man with new responsibilities began.

Back at the squadron nothing changed; flying duties continued as before. However, at the end of the duty-day, instead of returning to my room at the BOQ that I had shared with Dave Hunt, I went home to my bride. We fit right in with the squadron social life, but now we were invited to events with the married couples.

Leaving VS-30

In October I received permission to take a cross-country flight to Denver to visit the Colorado School of Mines. It was a special request because it was much longer (over 1,500 miles) than most approved cross-country flights; Norfolk-based squadrons were usually confined to destinations east of the Mississippi River. Jay Woodbury agreed to come along as my copilot. Our flight required two refueling stops at Wright Paterson AFB, Ohio, and NAS Hutchinson, Kansas, then on to NAS Buckley Field east of Denver. I rented a car and we drove to Golden, Mines' hometown in the foothills west of Denver.

Without any appointment I went to the Mines' Geophysics Department and was able to talk to the department head, Dr. John Hollister. He was very gracious and impressed that I had taken the trouble to fly out to visit the campus. I told him of the recommendation I had received from my Columbia profs, and he agreed that taking graduate courses at an institution different from that attended for undergraduate work could be beneficial. I asked if it would be possible to enroll at Mines for the semester beginning in January. He said I would be welcome. Objective achieved, Jay and I flew back to Norfolk.

After the trip to Denver I decided to interview with Eastern Airlines to judge whether flying for an airline might be a better alternative. My interview with Eastern went well, and I was told that I could expect to be offered a job when I left the Navy. Now Pat and I would have to make a decision as to which was the best course of action. Flying for Eastern would mean an immediate paycheck. Going to graduate school was a much more uncertain career path, and we would be living off the G.I. Bill, not exactly conducive to a life of ease. Pat, a University of Connecticut graduate, worked as a secretary when we were dating. She would get a job if we went to Mines, which would make life a little easier as I worked toward a degree. Actually, I imagined being a commercial pilot might become boring. We decided on graduate school and made our plans to leave.

The squadron threw a farewell party for us in December. I received the standard gift, a pewter beer mug engraved with the squadron's insignia and my name. My last S2F flight was December 29, 2.8 hours, all logged as copilot time. I might as well have been a tourist, a rather sad ending to my time in the squadron, but I left with many great memories. I was released from active duty on January 11, 1956. We packed the Oldsmobile again and returned to Pat's family home for a few days before heading out to Colorado..

Serving in the military as an officer is a very rewarding career. In addition to knowing that I was fulfilling an important duty as a Navy carrier pilot, a difficult job that few can say they are qualified to perform, I made friends to last a lifetime. My reasons for leaving the Navy are detailed above, but I must admit that until the end I wasn't sure I was making the best choice. Now, in retrospect, I believe I did. My view of the world, and my desire to be as creative as possible in everything I attempted, really didn't fit the more structured military life. However, I still attend VS-30 reunions every two years and stay in touch with several of my former squadron mates. Flying provides a special bond for those who take to the sky. The freedom one feels when commanding a powerful airplane is probably impossible for others who have never experienced that thrill to understand.

When friends ask about my flying experience I always respond with feigned humility, "There are two kinds of pilots, those who land on aircraft carriers and those who don't." Carrier landings were the most exciting and challenging operation I ever experienced, especially at night. Now, as I sit in my recliner thinking back 60 years, the memories of those landings are still vivid and I can partly relive them with the movies I made while serving onboard the carriers.

USS Mindoro CVE-120 anchored at Gibraltar with VS-30 on board, June 1954

AF Guardian - two versions, Killer above Hunter

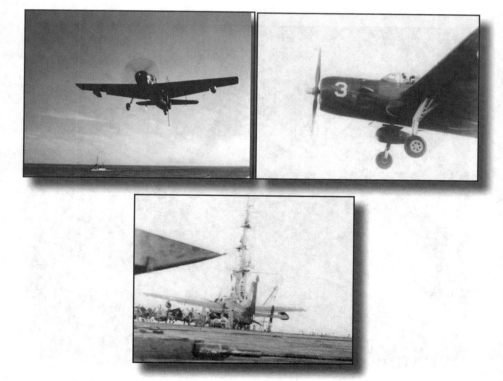

In the groove-cut-caught a wire. Nothin to it. Pictures of my landing taken by Bob Kowalski from Mindoro LSO platform

1954 Oldsmobile - not the original

Wedding party, August 1955

With Harvey at Bobby's home, Christmas 1954

Chapter VI
Graduate School - Colombia, S.A.
1956 to 1963

Colorado School of Mines

We left Stamford on January 15 and arrived in Golden three days later. Although home to the Colorado School of Mines, Golden, a former hub for 19th century miners going west to find their fortune, was not very large. It had a few motels and one hotel; we stayed in a motel at the north end of the city. There had already been some heavy snowfalls and snow was piled in high mounds in the center of the wide main street. I went to the administration building to register for the next semester's classes and get help in finding an apartment. We were directed to a house on 17th Street where a widow was offering a small furnished basement apartment. It was only two blocks from the campus and Berthoud Hall, where many of my classes would be held. The rental price was good and the owner appeared to be a nice woman in her early sixties, accustomed to renting to students.

Living in a basement apartment with the landlady overhead was a new experience for both of us, but we were able to ignore her constant presence and every footstep she took. The apartment was very basic, with linoleum laid on a concrete floor. The one bedroom had a squeaky double bed, bureau and curtained-off closet. The kitchen was very small, with stove, refrigerator, and a sink that doubled for morning shaving. In the corner was a shower stall behind a plastic curtain. The bathroom was just a commode. The main room contained a sofa and two chairs and a small dining table with chairs. One of our purchases in Norfolk had been a Sears 18-inch TV complete with rabbit ears, which we placed on a small table near the sofa. Reception in the basement was terrible, and there were only a few stations even if it had been good. The apartment had a private entrance from the driveway where we parked the Oldsmobile.

Pat quickly found a job as secretary for the Athletic Department. Fritz Brenecke, a well liked and genial fellow, was department chair and head football coach. The staff was small and included four other coaches. Mines did not expect its athletic teams to attract students, but it sponsored teams in almost every sport. All the staff coached two or more sports, aided in some by faculty. The athletic department office was a few blocks away from our apartment in the field house. They were a friendly group of guys, two about my age, and we got along well together; sometimes I played basketball with them. Pat held the job until we left, and it was a welcome addition to our very modest income.

Across 17th Street was a small housing development of WW II wooden barracks, each converted into a duplex apartment. This was subsidized housing for Mines' married couples, many of whom were returning to college after service in the Korean War. We had considered an apartment there but thought the basement would provide more privacy. Many new friends lived in the units, and it was convenient to meet there for the occasional social event.

I do mean the occasional social event. Classes were not for lazy students. For the next two and a half years I usually worked seven days a week, often past midnight. Although Mines was a state school that was required to accept Colorado high school graduates with acceptable SATs, its reputation for toughness was well known. As a result, it attracted only those Colorado students who were prepared to work hard and had strong science and math backgrounds as well as students from all over the US and many foreign countries. When I started there was only one woman undergraduate who wore heavy boots and jeans like the guys; there were no women in my graduate classes. Mines graduates were highly recruited by companies engaged in mining and petroleum exploration and exploitation, as well as a variety of other private sector companies and government agencies that required new hires to have strong engineering backgrounds.

Just as men from the Old West carried leather holsters on their hips, Mines students carried leather holsters on theirs. The difference: students carried slide rules in the holster instead of six-guns. The Christmas before Pat and I moved to Colorado my father had given me a Keuffel & Esser log-log-duplex-vector slide rule with my name embossed on the leather case. I'm not sure what it cost, but I expect it was over $100. For those without knowledge of what a slide rule is, they were invented in the 17th century and are devices that permit the user to make complicated calculations rapidly by moving the rule's middle slide and reading the required matching number on one of the appropriate body scales. Solutions to problems using simple math or log tables and different constants could be calculated as fast as you could move the middle slide. At Mines, knowing how to use a slide rule was essential in engineering courses to arrive at answers to exam problems. Mine was a 1956 substitute for the still to be invented laptop computer or digital calculator, perhaps working faster for some types of calculations than those devices when they are used today. I quickly learned how to use my slide rule.

Graduate School Studies

For my first semester I enrolled in five undergraduate courses which I had not taken at Columbia but were required for an undergraduate degree in geophysics at Mines as well as a graduate degree. I wanted to get them out of the way before tackling course work at the graduate level. It was tough! Getting in the swing of class lectures, labs and homework after more than five years away from college meant a lot of late hours. Even more difficult, I was competing for grades against younger students who had already survived two years of the Mines grind. The result: two C's, two B's, and one A. I could tell it wouldn't be a walk in the park, but by the end of the first semester I could plot out the road to a degree.

I took one of the most important undergraduate courses during the summer of 1956. It consisted of field trips and mapping the Wild Horse Anticline southeast of Pueblo. Mines had a permanent camp set up there; all geology majors normally took the course at the end of the sophomore year. I reported to the camp in June along with about 40 undergrads. As I remember, I was the only graduate student.

The first part of the course consisted of studying geological features at locations all over Colorado with lectures provided by various members of the faculty. After these visits we prepared for our final exam, mapping the anticline. This was done in two-man teams. I volunteered to team with a Brazilian student whom I had befriended, Carlos Campos, who was struggling with English. Although an undergraduate he was a year or two older than I, so we were the oldest guys in camp. Using my poor college Spanish I was able to communicate with Carlos by having him translate my Spanish to Portuguese and then English. Off we went. We were given four days to map the anticline. We completed our map and received an "A" because we were one of the few teams to map a minor fault that intersected the north rim. If you didn't recognize and map that fault you wouldn't receive an "A" for the course.

Carlos never received a degree from Mines. However, he went back to Brazil and eventually became the head of Petrobras, the Brazilian national oil company. I learned this when I read that Brazil had discovered a large offshore oil field and named it Campos Field. I called a Mines graduate in Brazil and he told me that it had been named after Carlos, my teammate in the summer of 1956. Unfortunately, Carlos had died shortly before I called and I never had the opportunity to talk to him again. I am sure he would have remembered that hot summer when we shared my water canteen while sitting on the outcrops.

At the end of the first semester my first action was to change my major from geophysics to engineering geology. It became clear that a master's degree in geophysics would require taking many more undergraduate courses, adding perhaps a year. My goal was to get a degree and find interesting work as soon as possible. Based on discussions with the geology department faculty, a thesis topic could be selected after the second semester in the fall of 1956. Over the next two years I took 66 semester hours of class work and audited another 20 or so hours of undergraduate courses I thought would be useful.

Joining the Navy Ready-Reserve

To further add to our income, and for some adventure to break the drudgery of my class work, in March I joined a reserve ASW squadron, VS-751, that drilled at Buckley Field. The squadron was still flying AFs, so it was nice to get back in a familiar cockpit. Three months later the squadron was decommissioned and I was able to join a reserve fighter squadron, VF-712, that also drilled at Buckley. Finally I would get my chance to fly a jet fighter, the F9F-6 Cougar.

I transitioned to the Cougar with four TV-2 flights. The TV-2 was a two-seat version of the Lockheed P-80 Shooting Star and was the jet trainer used at NAS Beeville when I went through advanced training at nearby Kingsville. It was a great airplane and ideal for learning to fly a jet. I was surprised to be able to solo in the Cougar after only four TV-2 flights, but procedures were a little more relaxed in reserve squadrons. I'm not sure how many flights were required to transition before soloing in jets at Beeville, but I would lay a small wager that it was more than four. On September 4 I took off for a short flight with another squadron pilot flying chase, a real thrill, flying solo faster than 300 knots for the first time.

By far my most exciting flights were those in September 1956 when the squadron went to NAS Miami for its two-week annual training cruise. One exercise involved intercepting a flight of B-47 bombers out over the Atlantic. In a four-plane formation we were vectored to the B-47s flying at 45,000 feet. We began simulated firing runs and discovered we were no match for the B-47s. When they saw us, they turned inside our approach. I sharpened my turn to stay on target, started to stall out, and during recovery ended in a steep dive. At 45.000 feet the underpowered Cougar was flying on the edge of its aerodynamic envelope. One short run was all we got and after I recovered from the stall the B-47s outran us.

To complete my VF-712 story, a flight one year later was really fun. To log more flying time, which was encouraged, I could schedule a flight almost anytime I had a few free hours. Then, not following any specific training exercise, I could fly around the area and sightsee. We were not supposed to fly over the Rocky Mountains west of Denver, but on a beautiful clear day I took a chance and flew out to the area where I had been doing my thesis. Having used aerial photographs to plot locations while doing field work, I was very familiar with how it looked from the air and easily found the Durham ranch where, as you will read, I had camped. Diving down, I buzzed the ranch house a few hundred feet above the roof, flew back up to 20,000 feet and took some pictures. Returning to the area a few weeks later, I stopped by the Durham ranch house to say hello. I mentioned that I had flown over, and they told me that when I roared over their house they thought there was going to be a crash and dove under the kitchen table, a questionable reward for my daring.

I continued flying with the squadron until December 1957, when I asked to be released to concentrate on my final classes and thesis. It was not until six years later, after returning to the States from Colombia, that I once again joined a Navy reserve squadron at Andrews Air Force Base in Maryland and then a flying billet at NAS Lakehurst, New Jersey.

Thesis

At the end of my second semester, in December 1956, I was ready to start my graduate thesis. My advisor was Dr. Peter Badgley, a Canadian. His specialty was structural geology, and I had taken one graduate course from him. He was not much older than I, and we had established a good relationship. Two other graduate students, Bobby Jean Kerr and Kenji Konishi (from Japan), were also ready to start their theses. Badgley suggested we map the structure and stratigraphy of three adjoining areas in western Colorado, an area Badgley was interested in for its oil and gas potential. Bobby, Kenji and I had taken classes together and had become friends. We thought this was a good idea as we would be able to help each other in some of the required field studies such as measuring the stratigraphic sections in each of our areas. Also, we decided to camp together as we would be in the field for some weeks and could share the housekeeping.

The area Badgley selected was south of Craig in Moffat and Routt Counties near Hamilton, named for a country store at the intersection of dirt and paved roads. We drove out together to get the lay of the land and find a place to set up camp. The three adjacent areas went from east of the Beaver Creek Anticline that included the Pagoda Dome, Bobby's area, west to the Iles Dome, Kenji's area. My area was in the middle and included the Beaver Creek Anticline. The Williams Fork River ran through Bobby's and my areas, while Kenji's area included a producing oil field.

We searched for a site in my area that would be the most convenient for all of us. We saw one off the dirt road that ran east out of Hamilton and appeared to satisfy that requirement. We introduced ourselves to the property owner, Harry Durham, and explained our need for a camp site. He graciously allowed us to set up camp on his cattle ranch in a grove of cottonwood trees a few hundred yards from his home. He also allowed us to take water from his well. The Durham family had a long history in western Colorado. We learned during occasional chats that, among other notable achievements, his great grandfather had carried the mail, often chased by Indians, from Hamilton to Rawlins. From there others would take it to connect with the Pony Express at Fort Casper. It was hard to believe that we would be tramping around on the same ground, 100 years later, minus the Indians, or that eventually I would buzz their ranch house in a jet plane.

Bobby had a tent large enough to fit three cots, and we were in business. This would be a new experience for Kenji, who was studying to be a paleontologist, as he had never camped out before. Bobby also supplied a Coleman stove and lantern, and with a folding table and chairs we completed our camp. We took turns cooking and buying supplies. Bobby had a pickup truck he could use, and I drove Kenji to his area every morning. We would then arrange for his pickup in the afternoon. After a few weeks in the field my brother, who lived in Grand Junction, lent me a Model A Ford pickup, the "Green Dragon," and I left the Oldsmobile in Golden for Pat to use.

The Green Dragon turned out to be very useful as it had high ground clearance, was narrow enough to drive through the many gates in the barbed wire fences I needed to cross, and got good gas mileage. Before we started we visited all the ranchers on whose property we would be working. Their only concern was that we close any gates we passed through to keep their cattle from getting out. We always did and never had any complaints from the ranchers.

We started in earnest at the beginning of July 1957 and worked until the end of August. I returned to Golden once for a few days. At the end of August, with all our field notes and samples in hand, we broke camp and returned to Mines. From that time until May 1958 I worked on my thesis and took additional course work. Pat and I drove out to my thesis area in January to check a location where I had collected a calcareous tufa, spring deposit. After reexamining the site, I concluded that the spring and tufa were associated with some small faults and added them to my structural maps.

By this time Pat was pregnant, but she continued working and typing my thesis at night. On April 24, with no complications during delivery, our son Thomas James (named after both his grandfathers) was born at Fitzsimmons Army Hospital, and a week later we brought him home.

I completed my thesis on May 20, submitted it for review, and prepared to defend it before the faculty review committee. "Geology of Part of Southeastern Moffat County, Colorado" included 104 pages of text and three appendices for a total of 174 pages. In addition it included three tables and five large colored plates illustrating different aspects of the area's geology and structure.

The largest geologic structure in my thesis area was the Beaver Creek Anticline. Almost all of the surrounding structures had been drilled with some success in finding either oil or gas. It appeared to be a very favorable site in which to find oil, and several wells had been drilled prior to my field work. All were dry holes. The Iles Dome in Kenji's area was a good-sized, producing oil field, and the Pagoda Dome in Bobby's area was beginning to produce gas. Why was there no oil or gas in the Beaver Creek Anticline? I proposed a theory to account for this "anomaly" based on regional uplift and faulting.

Returning many years later, I visited the Durham ranch. Not far from their home, where I had mapped the faults associated with calcareous tufa on the south flank of the anticline, I saw several oil wells pumping away. As our theses are all filed at the Mines library, they are available to anyone with the time to read them. I went back to the library to find out who had checked out mine, and there it was: an oil company had read it several years after it was filed and now had producing wells in the small fault block I had mapped.

Job Offer

In anticipation of graduation Pat and I had been mulling over where we should look for a job. I was leaning toward oil companies with operations in Indonesia or South America, the latter being more acceptable because of my Spanish language background. I reasoned that foreign employment would provide the best opportunity, as my fellow graduate students showed no interest in working overseas. Also, it appeared that overseas work paid the most and provided other benefits. Besides, oil exploration and production in the US were old hat; I expected there would be no really exciting assignments. To test the waters I had written to several companies expressing interest in working for them. All responded politely but with no specific job offer.

For one of my graduate seminars I had researched the oil industry in Colombia, S.A., and presented a paper on the results. It was a fascinating study because among other findings there was a report on one of

Colombia's first big oil discoveries near Tibú, a small town close to the border with Venezuela. In 1933 Petrólea-1 was spudded in on a geologic structure thought to have potential as an oil field. Shortly after drilling began, as the drill reached a depth of only a few hundred feet, the well blew out and an uncontrolled gusher of oil erupted, flowing at a rate estimated to be 10,000 barrels a day. The well caught fire, and it took 47 days to plug it and stop the flow. That story was enough to make any oil man's ears perk up, another Texas Spindletop! Based on the information I could find, for many reasons including major changes in the political environment, Colombia in 1958 was beginning to reap the results of decades of oil exploration. It sounded like an interesting place for a petroleum geologist, which is what I aspired to be.

While I was finishing course work for my master's degree and preparing to defend my thesis, Mines held its annual job fair. All the major oil and exploration companies and many smaller companies involved in activities related to the earth sciences and engineering came to recruit new employees. Mines was considered a very fertile ground for recruiters, as its graduates had a well deserved reputation as being hard-working and well grounded academically.

This was the first Mines job fair I attended. I walked into the auditorium with my thick thesis under my arm, prepared to discuss it and provide a graphic demonstration of my knowledge. A number of tables were set up, each with a company name and a recruiter standing by with literature describing his company. I was immediately drawn to a table with a placard reading Mobil Oil de Colombia. How about that for a coincidence! Behind the table was a handsome young man whose name tag read Carl Spalding. We began talking, and I learned he was the exploration manager of Mobil's Colombian operation. He was at Mines in the hope of hiring an exploration geologist and a petroleum engineer. He seemed surprised at my knowledge of Colombia's geology and petroleum industry and my obvious interest in working there. He offered me a job on the spot! Apparently he didn't find a petroleum engineer, as a Mines graduate never showed up while I was working in Colombia for Mobil.

Now, with a job offer on my resume, I finished my studies and successfully defended my thesis. One of the profs on the board suggested that, instead of going to work for Mobil, I spend another year taking a few more courses and upgrading my thesis to qualify at a Ph.D. level. It was interesting that he thought I could do it in only one year, but his suggestion was not very appealing and I declined. Actually, I think he was hoping I would agree and provide a low-cost graduate assistant to help teach his courses. But we were tired of living on a subsistence income in a small basement apartment and, with a new son, I was ready to get on with what I believed would be a challenging career. I thanked them all for the opportunity to study under their tutelage, and we prepared to leave.

After the graduation ceremony we said goodbye to our friends, shipped a few belongings to Pat's parents' home, packed up the Oldsmobile, and drove to Stamford. From there I commuted to Mobil's headquarters in mid-town Manhattan. I took a physical; signed all the paperwork; received inoculations for yellow fever, typhoid, typhus and tetanus; applied for a passport and work visa; and then received my ticket to Bogotá on Pan Am. All this was done in two weeks! Mobil was very well organized. Our plan was that after arriving in Colombia I would find a suitable house and then send for Pat and Tommy. With both the right and left arms still sore from the shots, my baggage containing work khakis and boots, I flew off first class on a Lockheed four-engine Super Constellation.

Oil Exploration in Colombia

On landing at Bogotá I was met at the airport by a Mobil agent, Manuel Arbalaez. He whisked me through customs (giving the agent a small gratuity) and drove me to the Hotel Tequendama, the newest and best in downtown Bogotá. For the next two weeks I reported to the Mobil office for orientation and meetings with the staff. Manuel was a jack of all trades with contacts everywhere in the government. He arranged for me to receive a pocket ID, with picture showing my affiliation, and a driver's license. In the meantime he drove me around the northern section of the city where most expats lived and showed me the type of housing I could rent at an affordable price.

The exchange rate with the peso was very good, so for those of us on the dollar economy rents were cheap. Based on my salary of $500 dollars a month, a grand sum compared to what I was used to, I settled on a three-bedroom row house with attached maid quarters and tiny patio enclosed in a high wall topped with embedded shards of broken glass. It was in an upscale, quiet neighborhood with a few multi-story apartments nearby. Across the street was the home of the editor of Bogotá's major newspaper, *El Tiempo*. His detached, high-walled home had 24/7 security, and he was chauffeured to work every day.

With this major decision out of the way I could send for Pat and Tommy. While waiting, Pat had gone on a buying spree and bought everything needed to furnish the house; we had never had our own furniture in Golden. It was packaged by a Mobil contractor in a large, sealed crate and shipped by boat to Barranquilla, Colombia's major port on the north coast. From there it would be trucked to Bogotá. Now I was ready to get out in the field and join my new partner, Jürgen Haffer, a German geologist whom I had yet to meet.

Before describing our field work, a brief geography and history lesson may be useful. Situated five degrees north of the equator, Colombia is a little less than twice the size of Texas. Within its borders are a complete range of environments from permanently snow-capped mountains towering 20,000 feet above sea level to dense virgin rain forest to tropical jungles bordering the major rivers, Caribbean and Pacific Ocean. When the Andes mountain chain enters Colombia in the south it splits up into three distinct branches, the eastern, central and western ranges. There is also a low mountain range bordering the Pacific Ocean and running north into Panama. In between the mountain ranges are wide river valleys. The largest, between the eastern and central ranges, is formed by the Magdalena River. It runs northward almost the entire length of the country, eventually emptying into the Caribbean. Bogotá, the capital, is located on a broad highland of the Eastern Cordillera at an elevation of over 8,600 feet. It was the capital of the Chibcha Indians before the arrival of the Spanish in 1536. Today, as then, because of its elevation and nearness to the equator, its residents enjoy a climate of perpetual spring.

Like many other South American countries, Colombia has had a troubled history. Its problems were fomented by strongmen and dictators who, through their greed, kept many of their fellow citizens in poverty. Colombia's last dictator, Rojas Pinilla, overthrew an elected government in 1953. In turn he was overthrown in 1957, the year before I arrived, by the military and a coalition of the Liberal and Conservative parties. Pinilla vacationed at times at a large finca on a hill east of Monteria in northwest Colombia where I would soon be working. The locals told stories, not very complimentary, about what happened when he visited with his security detail.

With his demise a long period of civil unrest known as La Violéncia ended after claiming some 250,000 Colombian lives. A compromise government (the National Front) was established in which the two parties agreed to alternate the office of president. Alberto Lleras Camargo became the first president under this agreement and was in office until 1962, when the office was peacefully exchanged.

Also, like almost all Latin American countries, Colombian society is highly stratified with a small upper class, a moderate-size middle class, and the greater majority living in near poverty. The good news, if there is any for most of Colombia's citizens, is that, because of its geographic location and mild climate, living is easy. The higher elevations require warm clothing at times, but most of its citizens need only enough clothing to satisfy modesty. It is possible to raise fruit and vegetables the year round, to have a few chickens and pigs, and to travel or carry heavy objects on the backs of donkeys and mules. Major infrastructure outside the big cities is neither required nor available.

After I left in 1963, communist organizations supported by Fidel Castro formed several guerilla groups aimed at changing the status quo. They specialized in terrorist attacks, kidnapping and demanding ransoms. These groups became drug kingdoms that have continued making violent attacks on Colombians, the vast majority of whom only hope to live peaceful, productive lives. As I write this many problems still exist and the US is supporting the central government in order to defeat the terrorists. But let's go back to 1958.

When I arrived, Colombia had begun a period of stable government. Foreign investment was encouraged to improve the economy and standard of living. All the major oil companies, Intercol (a subsidiary of ESSO), Shell, Texaco, Cities Service, Sinclair and British Petroleum, as well as Mobil, had obtained large concessions in which to explore for oil. In addition, the national oil company Ecopetrol and smaller oil and service companies all had offices in Bogotá and participated in oil exploration and production.

Government land grants (concessions) were leased to private companies to encourage exploration that would hopefully lead to oil or mineral production. To obtain a concession a company had to make a down payment and promise to conduct a set amount of exploration in a given period. If the results were negative, the concession would be given back. However, if a company wanted to keep the concession after the exploration period, it had to drill wells or open mines. This was a reasonable way to do business. Once it was clear that political stability had returned there was a scramble to obtain concessions in areas that appeared to have the best prospects, although in many cases it was just a guess as to which were the best areas for lack of good data.

Huge areas were now under contract. For these reasons Mobil was anxious to get geological and geophysical field parties working to determine which concessions to keep. If the exploration phase provided encouragement, the next step, drilling wells, entailed a commitment to very large investments. A typical wildcat well would cost $2 million dollars or more, with no assured return on investment if it turned out to be a dry hole. Reports of the results of all the work had to be filed with the government so that it could determine if the required conditions were being met.

Mobil, along with COLPET, a 50-50 partnership with Texaco, had many concessions in northwestern Colombia as well as some in the area extending from the Magdalena River east to the Venezuelan border. These latter concessions were now producing over 100,000 barrels a day from the oil fields I described in my graduate seminar paper. When I arrived oil production was on the rise, making Colombia one of the world's largest oil exporters. Oil produced in the eastern concessions was carried by pipeline to Coveñas, a tank farm and port on the Caribbean, where tankers were filled with the crude oil. Built in 1939, the pipeline is considered one of the world's engineering marvels. It traverses some 250 miles of jungle and swamp, crossing parts of the eastern Cordillera and the Magdalena River before running the last 65 miles in a straight line from Mompós to Coveñas.

In the easternmost section, before the pipeline crossed the mountains, surveyors and construction crews not only fought off mosquitos and other tropical pests but were occasionally attacked by the Motilone Indians, a tribe that never made peace with the Spanish or those who came later. When flying from the Mobil concession in Tibú to Coveñas, company pilots in Mobil's DC-3 would circle over their settlements so we could take pictures of the large communal huts set in jungle clearings. In 1958 the tribe was still considered "bravo," meaning one would risk his life entering their territory.

Field Work Begins

With a home rented, and Pat and Tommy soon to arrive, I left Bogotá on an Avianca flight to Monteria in northwest Colombia. I was met at the airport by Jürgen and some of the Colombian crew, including the foreman (capatáz) Alfonso Figueroa and truck driver Rosembérg Paternina. All were very friendly, and I am sure Jürgen was happy to see someone with whom to share the work. We spent the next few days in the Monteria area and Jürgen took me to several sites where outcrops were exposed so that I had an idea of the local geology and how to identify the geological formations we would be mapping.

To visit outcrops on the west side of the mighty Sinú river, almost a mile wide at Monteria, I was introduced to a new mode of transportation, a ferry without a motor. The ferry, only large enough to carry a few vehicles plus assorted other passengers including horses, mules, donkeys and people, was attached by a single-strand wire to a pole anchored in the middle of the river about three or four hundred yards upstream. I don't know how it was anchored, but it must have been pretty substantial as the ferry was very heavy and in use all day long. It was the only way to cross the Sinú; there were no bridges. After loading, the ferry master pushed off from the bank and, using a large hand-steered rudder, let the current push the ferry across at a very good clip. It was simple and effective, and the ferry operator never had to buy fuel. According to stories told, every once in a while the wire broke and the whole contraption ended up downstream, cutting off the city from the ranch land on the west side until the ferry was back in operation. But in the five years that I used it I never encountered a problem.

After this brief introduction to northern Colombian geology we drove south in Mobil's Dodge Power Wagon on a gravel road that paralleled the Rio Sinú, passing large ranches and open range. The ranchers raised Cebu, a type of cattle originally from India that could tolerate the heat and insects of tropical climes. At the little town of Tierra Alta we crossed the Sinú on another motorless ferry and turned southwest on a dirt track. Eventually that disappeared, and a short time later we stopped where two more of the crew were waiting with mules. This would be my first adventure riding a mule, or a horse for that matter, but just the start of logging hundreds of miles in the years to come. As we rode along I used my rusty Spanish to ask questions of the crew, who spoke only Spanish. Jürgen's Spanish was quite good and his English perfect. If I had difficulty, I would repeat my question in English and he would translate and clear up any confusion. An hour later we arrived at the camp, a small tent for Jürgen and three large tarpaulins covering the kitchen, crew hammocks and equipment. I was shown to my quarters, a cot enclosed by a zippered mosquito net under a small waterproof cover.

Jürgen had been in Colombia for the past eight months. He joined Mobil in November 1957 and two months later went to Monteria with Bill Lawson, the Mobil geologist responsible for conducting field work in the area. Lawson was scheduled to be transferred to Iran, where Mobil was beginning to work. Although there were reports available covering Lawson's work, it was important before he left to transfer his knowledge first-hand to Jürgen. This meant visiting many important outcrops where the subsurface formations came to the surface and could be studied, including a few I had just been introduced to. It was a crash course, and Jürgen was now putting it to use in the field.

In the next days we continued the surveys Jürgen had begun before I arrived. We walked or rode on our mules, mine a gentle white female called Nora. We stopped whenever we could find an outcrop to take measurements. Returning to camp in the afternoon, we drew up the day's traverses, checked the location of our sampling stops on aerial photos, and described what we had observed in our field books. After a few days I no longer felt the pain in my legs and buttocks from sitting in a saddle for five or six hours each day. We used western saddles and the mules were very broad, stretching my hip and knee joints in ways to which they were not accustomed.

The geology was very interesting, although I did not yet have a good picture of the sequence and characteristics of the geologic formations. But everything I saw was so new that it was enjoyable just to soak up the atmosphere. Riding slowly on a mule allows one to contemplate his surroundings with a different and deeper awareness of the world than the normal fast-paced life provides. In the evenings Jürgen and I spent time getting acquainted. Living in close proximity 24 hours a day with someone I had just met required a period of adjustment. But we hit it off from the beginning, and our friendship continued until his death in 2010.

I returned to Bogotá after a few weeks in the field in time for Pat's arrival with Tommy. They stayed in the Tequendama awaiting our new furniture. Once they were settled in at the hotel I rejoined Jürgen. We worked until August when the field season drew to a close, with afternoon showers restricting the time we could spend in the field each day.

We had one diversion that occupied us for an hour or so each afternoon. After finishing recording the day's surveys we caught butterflies. Jürgen's father, a teacher, was a butterfly collector and had written his doctoral thesis on caterpillar nerve systems. Being in Colombia gave Jürgen a chance to add to his father's collection with new tropical species. Before I arrived he had been collecting what could be found near his campsites. I joined in the hunt, using his butterfly net. I would catch a few every day, and Jürgen would carefully place them in glassine envelopes.

At a few camp sites where an isolated settler had a hut nearby, young children might come in the evening, attracted by our Coleman lamps, for their first exposure to "modern" civilization. They stood silently outside our mosquito-net tent watching wide-eyed as we ate. Strange people, these fair-skinned foreigners. Using my rudimentary sketching skills I would ask them to pose, draw their faces on a page from my notebook, and give the sketches to them. They usually laughed and ran back to show them to their families. I wish I had had a Polaroid camera, which would have been a real hit.

One of the most spectacular sights of all my years in Colombia was experienced when we were surveying the Rio Naín area, south of Tierra Alta. In a tributary of the river, the Quebrada Resbalosa, the water was clear and very shallow as it was still the dry season. It was overarched by huge rainforest trees and mostly in shadow. Occasional sunbeams would penetrate the dense canopy and sparkle on the water. Coming around a bend on our mules we surprised a large colony of butterflies drinking in the shallow water, the *Morpho iris*. A large butterfly with a wingspan of six or seven inches, it is one of the most famous tropical butterflies because of its spectacular coloring. The top of the wings is an iridescent deep blue, the underside a light tan and non-reflective. As the butterflies took off down the Quebrada ahead of us we saw dozens of alternating flashes of bright blue, like strobe lights going on and off, as they opened and closed their wings and flew in and out of the sunbeams. We collected a few which, along with the rest of his collection, Jürgen planned on mailing back to Germany when he returned to Bogotá.

The reader may ask how one carries out geological exploration and mapping in a country where little is known of the geology and there is almost no infrastructure to provide support. In Colombia it meant working six or seven months of every year during the dry season. We used a Brunton compass and rangefinder to record our location and geology, supplemented with aerial photos when available, returning to Bogotá during the rainy season to write reports. Other assignments might also come up, such as working as a well-site geologist during oil well drilling, that I will describe later.

Although a few small villages were widely scattered in our field work areas, most of our studies took place far off the beaten path. The only way to work productively was to live in small, easily moved camps and travel and transport all our equipment by mule. Mobil owned 20 pack and riding mules. Where we worked only rough trails and a few dirt roads crisscrossed low, mountainous terrain, and in some cases there were no trails at all. Then our crew, with their machetes, cut their way through the undergrowth. Much of the region was still covered with virgin rain forest although, even as early as 1958, clearing was proceeding rapidly.

With the dry season almost at its end I had a lot of catching up to do in order to be useful. My graduate thesis was based on studying and deciphering the geology of a small part of western Colorado, where at every turn in the road the Earth's geological secrets lay open and exposed to the practiced eye. Conducting geologic studies in open savanna and tropical rain forest required a much different approach and, to some degree, a re-education. Surface formations were poorly exposed and usually deeply weathered and eroded. It was difficult to find fresh exposures from which to make measurements and collect samples.

Where possible we conducted our studies by wading in streams and rivers. During the dry season the low water levels gave us the best chance of finding exposed formations. A good, clean outcrop was like a gold vein to a prospector in the old West. Finding one would make our day, since we could be pretty sure that the information collected would be accurate and allow us to correlate our measurements with previous observations. A fresh outcrop also meant we might find a few microfossils to return to the paleontologists in Bogotá, who could then accurately date the formations.

Our crew consisted of eight or nine Colombian workers with various skills, all of whom came from the vicinity of Monteria. The most important was our "super capatáz de la selva," Alfonso Figueroa, who was with the greeting party at the airport. He had worked for Mobil for over 20 years. He and our Monteria agent Jose Marmol, who purchased food and equipment when requested and forwarded mail or other information, were the only employees paid year-round. During the rainy season, when we were in Bogotá, Figueroa was responsible for maintaining and repairing our equipment and looking after the mules which were kept in local pastures. At the end of the rainy season he would recruit workers for the new season, most of whom were rehired from the previous year, to be ready when we returned in January as the dry season began.

Camp sites usually consisted of a small area cleared down to bare earth so that we would be free of all critters, large and small, that called the forest or savanna home. There were four large tents - one for the cook, his hammock and all the kitchen equipment; another for the workers' hammocks; a third for our saddles and mule accessories, and finally our tent, a large mosquito net enclosure strung beneath an overhanging tarpaulin. The first camp described earlier was not typical, as Jürgen had been working alone and I was with him for only a few weeks; he didn't need the big mosquito net tent.

Within our mosquito net tent were two cots and two folding tables for writing our reports, drawing our maps, and eating. Other "furniture" consisted of canvas-backed folding chairs and made-on-site poles to hold our clothes and equipment above the usually damp ground. On a few surveys we used the small tent that was just big enough to hold our cots. Occasionally, if a short side trip was required to study a special feature, we took just enough equipment and supplies for two days and slept overnight in hammocks under a tarpaulin.

Before leaving the US I bought a Zenith Transoceanic battery-operated radio. It had a place of honor on one of the tables since it was in almost constant use when we were in camp. We could pick up Voice of America, Armed Forces Radio and, crystal clear, Cuban radio stations. No wonder Cuba was so successful exporting revolution to many Latin American countries; it was the only Spanish-speaking station we could pick up even in the most remote area. If we were working near a road, we would use a small electric generator. Usually, however, we were too far away to carry the heavy generator and fuel, so we used Coleman lamps.

Figueroa was a natural leader, tall by Colombian standards (about 5' 9"), soft spoken, with a classic indian profile of high cheekbones and a slightly hooked nose. He could perform all of the necessary jobs better than any of the workers he hired. He was an expert "woodsman" who was comfortable working in the rain forest or in the surrounding mountains and plains that stretched from the Magdalena River to Panamá, an area about the size of South Carolina. Having worked with many other Mobil geology field parties, he understood what we needed to accomplish and could anticipate our requirements. He was an indispensable member of our crew, and we depended upon him to keep our small team productive and running smoothly.

In addition to Figueroa we hired two mule men, a cook, and four or five helpers who did all the other work from cutting trails to setting up camp. As mentioned, we did our mapping with a Brunton compass and rangefinder, recording the readings in a field notebook. We became very good at using this rudimentary method and could usually close a traverse of several miles within a few hundred feet. Members of the crew not cutting trail or walking the mules assisted by carrying our homemade surveying rod, made daily from a small straight tree with the bark at the top shaved to a white core, and sample packs.

Our mule men, Manuél Montez and Julio Caro, were very wise in the ways of mules and were rehired every year. Mules are much smarter than horses, so the men had to stay ahead of what the mules were thinking. Sometimes we were in an area that had fenced pastures where the mules would be put after we finished work. This did not assure that they would be there in the morning. My mule, Nora, had learned how to open pasture gates using her teeth to lift the wire strap; I have a home movie showing her doing this. Thus, on more than one occasion when in the morning Manuél and Julio went to round up the mules for the work day they were nowhere to be found. Nora had opened the gate during the night. The men would then track them in the direction of our last camp and find them, sometimes miles away. On a day when this occurred we might be delayed several hours before we could leave camp. On the occasional trail that was closed off with a gate I did not have to dismount; Nora opened the gate and someone following closed it. Usually the mules were tethered near camp at night and fresh cut grass and leaves were placed where they could reach them. When possible we carried corn to supplement their diet.

Here's a little mule lore for the uninitiated. Mules are male or female; however, they are sterile and cannot breed. Like donkeys, mules take dirt baths by rolling on the ground. One can tell that he is buying a strong mule if, when taking its dirt bath, it can roll over its spine to both sides without getting up. Mules are very sure-footed and will not step or place themselves in dangerous positions. How they foresee such dangers only a mule can tell, and even clever Nora never told. Knowing this provided a comfortable feeling when riding on slippery mountain trails. Nora was very large and strong but was exclusively for riding; she was never used as a pack mule except to carry me and my equipment. She had been trained to run at a very comfortable gait, barely lifting her hooves above the ground. During the infrequent times when we moved over level terrain and could travel at a gallop I would ride along smoothly, never bouncing on the saddle.

After a day's work Manuél and Julio washed down the mules if our camp was near a stream and checked our saddles to see if they needed repair. In several areas where we worked vampire bats flew at night, and in the morning the mules would have bleeding wounds on their necks. This never seemed to bother them, and the blood was washed off before we mounted.

On camp moving day pack saddles were used to carry all our equipment and only Jürgen, Figueroa and I would ride; the rest of the crew walked along keeping the mules going. As we usually were extending our surveys along a stream or river, we leapfrogged the camp in the direction our survey was advancing for a distance we estimated would be several days of work beyond where we last ended. On the first day from the new camp site we would go back and start at the last point and work toward camp. Then we worked several days beyond the camp site. By doing this we had to move the camp only once a week. This was the most efficient way to use our time because moving camp cost a day of surveying.

Our cook, Rafael Sierra, a man with a sunny disposition and a perpetual smile, was the second most important crew member. He had been rehired many times and managed to keep us all well fed with only primitive kitchen equipment. He cooked on two Coleman stoves using fuel carried with our food and equipment on the backs of our mules. Without refrigeration and eating mostly packaged and canned food supplemented with rice, local produce and fresh game when available, we never had any food-related problems during my five years in the bush. As a dessert lover I suffered the first two years. While on our first vacation back in the States I bought a small "ovenette" that fit on top of the Coleman stove. Sierra could then bake a pie or cake when he had the proper ingredients.

Every morning he also baked a non-yeast bread that he served as toast with hot chocolate made of powdered milk and the additive Milo. Why hot chocolate and usually hot oatmeal for breakfast? Because, believe it or not, the early mornings were usually quite cool, especially when working in the mountains, and we had just spent the night under a blanket in an open tent. Our meals were not exactly five-star but certainly better than one might have expected so far from civilization. One of my favorites was vegetable soup made with chicken livers when Sierra could buy some chickens. His kitchen was as spotless as possible under such conditions, and he had the good habit of washing his hands between chores, a practice I am sure contributed to our good health.

Getting a supply of clean drinking water was always difficult. One of the earlier Mobil geologists had discovered a Swedish system that utilized a ceramic filter to purify water. The local stream water in a pail would be pumped under pressure through the filter, which was advertised to remove contaminants, especially amoebas. Sierra would fill our canteens each day and the system worked well; we very seldom suffered from stomach ailments or other illness associated with bad water.

Sierra had another important job, making sure we had clean clothes. We wore long-sleeved khaki shirts and pants tucked into our socks and boots to keep out the ever-present insects. At the end of a day of wading in water and walking muddy stream banks we returned to camp dirty and sweaty. Sierra made sure we had clean clothes the next day and oiled our boots every evening. Often he was able to find some local ladies to do the wash. On returning to camp in the afternoon we would find him entertaining the women in the kitchen tent, showing off his "modern" appliances and cooking prowess before a rapt audience accustomed to cooking over open fires. We gave him a small fund to pay for these services as well as to buy any local produce available, so he must have appeared a very wealthy person to these poor homesteaders.

How or where he was able to find these willing workers on short notice, even in the most remote areas with no signs of habitation, we were never able to figure out. He would somehow get the message out that he was hiring. His reputation as a ladies' man with our other workers did not suffer when he demonstrated this great talent.

Our bathroom facilities, as one might expect, were very primitive. We dug a hole about 50 feet downwind from the camp, placed a canvas on poles around three sides for privacy, and squatted. When we broke camp we filled the hole back up. If we had to use the "bathroom" at night, we carried a flashlight because tarantulas and snakes came out of their holes and would sometimes be seen on the path. This was not a big deal, as they were more interested in getting out of the way than attacking our bare legs.

We showered by standing in a nearby stream or river pouring water over our heads from a cup. We used a blue mercury-based soap, the one that is used by staff in hospital operating rooms, and this successfully kept crawly things from hiding in body crevices. We took a quinine tablet each week and made the crew do the same. Most or all of the crew already had malaria, so this was the best way to keep them from having an attack and being confined to their hammocks until it passed. We carried snake-bite kits but never had to use them although we often encountered poisonous snakes and at times large boa constrictors.

Twice we had run-ins with army ants that moved at night through our camp sites. Both times our tent was in their path; we just got out of the way and let them march along in their tightly packed column. In a half hour or so they were gone and we went back to sleep.

As mentioned, none of the crew spoke English, so I had to rely on the Spanish I took in high school and college. Those courses were rote learning of grammar and vocabulary; they didn't emphasize conversation. But with this ten-year-old background I was able to understand the local dialect and to give instructions and chat with the men. As my time in Colombia increased, my Spanish improved considerably. Those living near the coast spoke an unrefined Spanish dialect, slurring some letters and syllables, as opposed to those living in Bogotá who prided themselves on speaking perfect Spanish.

We carried several hundred dollars in pesos to pay for any necessary services, such as overnight pasturage for our mules if we could find any local landowners. We paid our workers every two weeks. This didn't require a lot of ready cash, as there was usually no place to spend it, and each worker would designate the bulk of his earnings to be given to his family in Monteria. At that time the dollar was worth about ten pesos and the trail cutters received about a dollar a day plus meals. The more skilled workers such as the mule men and cook were paid more. Figueroa, our highest paid worker, received almost two dollars a day. We accomplished the payment transfers to the crew's families by meeting with Paternina at prearranged times and places; he would carry the messages back to Jose Marmol in Monteria, who would then disperse the amount agreed upon from his funds.

Most of the crew kept some of their earnings in our "bank," an account we kept that would reflect what they earned minus any disbursements. At the end of the field season we paid them the lump sum they had saved plus a bonus, and they returned home with a sizable (for a Colombian worker) nest egg to tide them over until next year. The wages we paid to all our workers were higher than those paid by local employers, so jobs with Mobil were eagerly sought. As a result we had the pick of the labor pool, and when we hired a worker who proved his worth we continued hiring him in succeeding years. With the above introduction, it is time to describe the most memorable exploration campaign in which I participated during my five-plus years in Colombia.

Exploring the Darién

In January 1959 Jurgen and I began our most difficult and exciting geological study. The area selected was on the west side of the Gulf of Urabá in northwestern Colombia, a region in which very little exploration had been done in past years. The Rio Atrato basin just south of the Gulf gets up to 390 inches of rain per year, making it one of the wettest areas on the planet. Aerial photography that was available during our 1958 studies and aided our work was missing in the area we would explore because of the almost constant cloud cover. Its absence would place a premium on our ability to carefully record the location of our surveys.

Based on our previous season's field work on the east side of the Gulf where Mobil held concessions, we believed surveys on the west side would add significantly to our understanding of the geology, especially of the older formations that might be drilling targets. Intercol was considering relinquishing its concessions east of the Gulf, and Mobil management was debating whether or not to pick them up. We hoped that subsurface formations on the east side would be uplifted and exposed on the opposite shore and eastern flank of the Tacarcuna mountain range that marked the border between Colombia and Panamá. If we found the formations exposed, then their potential as oil sources and reservoirs could be identified. Sun Oil Company had already begun drilling a well on the east side near the town of Necoclí. The overall area was of great interest, especially if the Necoclí well discovered oil.

A little history and geography may spice up the story. In 1510 Vasco Nuñez de Balboa established the first city, Santa María de la Antigua, on Tierra Firma, the South American continent, as opposed to all the cities the Spanish had built on islands in the Caribbean. Based on the description found in the *Spanish Chronicles*, the site Balboa selected was on the west side of the Gulf of Urabá, just north of the Rio Atrato delta. It appears that the city was located near or on the Tanela River in present-day Colombia, but the exact location can't be determined from the sketchy description. Regardless, it was a poor choice. Once the largest city on the mainland of the newly discovered continent, it is believed to have housed at its peak as many as 5,000 settlers, bureaucrats, nobility, soldiers and indians. However, even in the dry season from January to March it might rain once a day for a short time. The constant rain, of which the settlers bitterly complained, was a contributing factor in abandoning the city in 1524. A new city, Acla, was established farther north in present-day Panamá.

Despite its poor location, according to the *Chronicles* that have survived, it was from Santa Maria de la Antigua that Balboa left on the expedition that resulted in his discovery of what we call the Pacific Ocean (so named later by Magellan). After subduing several indian tribes farther north, Balboa learned from the son of a local chief about a vast sea to the west. Again, according to the *Chronicles*, on the morning of September 25, 1513, with a select group of 80 men, he left the indian village of Quareca in the Panamanian province of Darién in search of the sea described by the chief's son. At one point, going ahead alone, he climbed a high hill and looked toward the southwest, becoming the first "Christian" to sight what he named the South Sea because of his direction of march.

The local indians also told Balboa of a people who lived to the south on the coast of this sea and were known to have vast amounts of gold. Searching for gold does not appear to have been a major preoccupation for Balboa, although he took whatever he could find when he subdued the local tribes. However, accompanying Balboa as a junior officer was Francisco Pizarro. For Pizarro, like most of the Spanish conquistadors, gold was the reason he had come to the New World. He determined to search for these rich cities to the south, and for this he needed boats. Material to build them was transported overland from the Caribbean coast to the Gulf of San Miguél, from which Pizarro began his voyage and conquest of the Inca. But Balboa never had the opportunity to profit from his expeditions. In 1519 he was tried for treason and beheaded at the command of his father-in-law, Pedro Arias de Avila. What little wealth he had accumulated was given to relatives in Spain.

In 1954 ex-King Leopold III of Belgium tried to retrace the route taken by Balboa when he sighted the South Sea in 1513. He hoped to begin his expedition, as had Balboa, from the city of Santa María de la Antigua. But as the city had been built of wood, palm fronds and clay, it had completely disappeared and he was unable to locate it. Almost certainly after the last Spanish inhabitants left the city the local tribes that had been abused by the Spaniards took anything of value that remained and then burned it to the ground. Leopold gave up the search and began his reconstruction of Balboa's trip from Panamá .

Today two indian tribes are descendants of those first encountered by Balboa, the Chocó and the Cuna. The Chocó are hunter-gatherers living mostly in the lowlands, moving from camp to camp depending on the season. The Cuna live on the flanks of the mountains in well-organized communities. Presumably because stories of abuse during the first contact with Europeans had been passed down through the centuries, the Cuna did not allow anyone to enter their territory until the mid-1930s. They were known to have killed those who tried. Because of their isolation and fierce resistance to any affiliation with Colombia, the Colombian government has allowed them to adopt a form of self rule and they now live quietly on their reservations. To the north the related San Blas Indians have adjusted better to modern civilization, and their island communities are tourist destinations.

During the 1940s and '50s, funded primarily by the US, the Pan American Highway was built. Now stretching some 16,000 miles from Prudhoe Bay in Alaska to Ushuaia in Tierra del Fuego, it winds its way through fourteen countries with occasional side roads branching off. However, there is one short stretch of about 68 miles that keeps the highway from being completely continuous over the total distance. One can travel by car from Alaska as far south as Yaviza in Panamá. There the road ends on the edge of mountainous rain forest. One must then backtrack about 100 miles to a port on the Panamanian coast and go by boat to Turbo in Colombia. From Turbo the highway continues to Tierra del Fuego. The highway still has many rough spots, but it is passable by four-wheel-drive vehicles except for this detour. A few adventurous spirits have completed the overland route including the 68 mile gap, but only with great difficulty by cutting their way through the swamps and rain forest and boating across the Atrato River.

The missing 68 miles are known as the Darién Gap. One would think that such a short stretch would have been easy to complete. However, after all these years the highway has never been completed for several reasons, the most important being the very high cost of crossing the often rain-swollen Atrato River basin. Jürgen and I would be exploring in this very area, including near the reported site of long abandoned Santa Maria de la Antigua.

Starting in September 1958 in Bogotá, we began to plan our work. Among other tasks was buying sufficient canned and packaged food to last the three months we would be gone, as we could not count on being resupplied once we landed on the western shore of the Gulf. We packed the petacas (heavy fiberboard cases that fit on the mules' pack saddles) with food carefully separated so that Sierra could select each day's rations. Canned goods included fruit juice, soup, fish and stew, plus powdered milk, crackers and cookies. We also expected to eat local food such as rice and fried bananas which were served to our crew. We did not expect to find any "grocery stores" in this remote region, essentially virgin rain forest except for a few very small towns and scattered settlers. We hoped that we would be able to shoot game and carried a twelve-gauge shotgun for this purpose as well as for bird collecting.

After the petacas were filled, we shipped them to Monteria on the company plane. Figueroa and Paternina would take them with other equipment in the Power Wagon via a long circuitous road to Medellin and then to Turbo, the jumping off point for our expedition. The mules and other workers would also meet us in Turbo, traveling overland from Monteria. An agent had been hired in Turbo at the end of the 1958 field season to help with all the arrangements.

At the beginning of January 1959 we flew to Turbo from Monteria by air taxi and found a small motel-like residence north of town where we could stay for a few days until everything was ready for the crossing. The motel was on the beach, and when we awoke the first morning we could clearly see the tops of the Tacarcuna Mountains some 40 miles away across the Gulf; some of the highest peaks reached to over 6,000 feet. We never saw the peaks again from Turbo as rain clouds over the mountains always obscured the view.

At the end of January we made our first attempt to cross to the western side of the Gulf. It almost ended in disaster. We had hired a small diesel-powered coastal freighter, about 75 feet long, to carry our 17 mules, supplies and ten passengers. By the time we had everything loaded and the mules secured, night was falling; work usually moves along slowly in Colombia.

Against our better judgment we decided to leave, as the local forecasters predicted the weather would be good. After all, this was the dry season. The captain assured us it was safe and better to leave now as the weather could change. We trusted him and shoved off. The total length of the trip was estimated to be about 50 miles, including the portion up the Atrato River, and should be completed in four or five hours. About one hour into the trip, while still in the middle of the Gulf, the winds began to howl, rain began to fall, and the waves grew higher. The boat began to roll precipitously and the engine quit. Now we were in real trouble. I went down below where the mules were tethered to see if help was needed. I could see the fear in their eyes as well as in those of Manuél and Julio as they tried to keep the mules from falling. Neither had been at sea before, and I am sure they thought we were lost. The captain finally got the engine running again and we directed him to return to Turbo. Back in Turbo's small protected harbor we tied up and off-loaded the mules and equipment.

After that experience we knew we couldn't depend on local transportation to get us across the Gulf. Jürgen called Bogotá and explained our predicament. Carl Spalding arranged for Mobil's LCU, a surplus WW II amphibious landing craft that was supporting drilling the Cordobá-1 well up the coast, to meet us at Turbo. Two days later it arrived and we started again. This time we were successful and arrived in the late afternoon at Sautatá, a former sugar plantation about 30 miles upstream from the Atrato's mouth, where we would begin our surveys. It was clear that, even if the weather had been good on the previous attempt, we would not have made it. Entering the Atrato's delta the LCU with its very shallow draft had to use its powerful engines to force a way through the silted-up and vegetation-choked entrance. It was only through the perseverance of the LCU's captain that we finally broke through to the clear, wide river itself. After that it was an easy trip up the river to the abandoned sugar plantation.

To shorten this story I will condense three months of work into a few paragraphs and highlights as we studied the geology south of Sautatá and then worked our way north to Acandí, some 65 miles away. While preparing for our exploration we found a study made south of Sautatá in 1946 but only fragmentary reports of the region to the north made in 1919, 1930 and 1942. ESSO's road map of Colombia, available at gas stations, showed a projected road from Sautatá to Acandí, but there was no evidence that even a survey had been done. With so little information to go by, we decided to traverse primarily the rivers and streams where we expected to find the best formation exposures. We would start at their mouths that emptied into the Atrato or the Gulf and go westward to their sources in the mountains. All the streams and rivers began in the mountains and ran in a general west to east direction. Before starting our surveys to Acandí we spent a few days studying the geology as far south as Rio Sucio and exposed along the Rio Cacarica.

Another brief history lesson: from the mid-nineteenth century until the early twentieth several southern alternative routes for a canal to connect the Caribbean to the Pacific Ocean were proposed and surveyed. Some used the Atrato River and went south on the river until meeting one of the many large rivers that emptied into it. Then the canal would go west along the chosen river. It was thought that only a short excavation would be needed to cut through the low mountain range separating the Pacific Ocean from the Atrato drainage. Some planners thought the passage might be possible without building locks despite the known difference in sea level between the Pacific and Caribbean. One of the routes surveyed by the US Navy Bureau of Navigation in 1882 started at the mouth of the Atrato River, went south to the Rio Cacarica, then west to the Pacific. For many reasons all these routes were discarded in favor of the present Panama Canal.

Once we started working north of Sautatá the scenery was, at times, stunning. I suspect most have in mind from movies they have seen that a jungle or tropical forest is choked with vegetation of all types. Not true. Only along riverbanks or the coast where sunlight easily reaches the ground do such conditions exist and make passage difficult. Because the canopies of huge trees like the ceiba shut off sunlight to the forest floor there is little undergrowth in the rain forest and one can move quite easily. These huge trees with their buttress root systems take up space, and thick vines stretch from the ground to the treetops, but usually we rode on our mules below the canopy without needing to cut a trail.

We saw many different birds, but other wildlife was scarce or stayed well hidden. Our mules and ten men made a lot of noise, so it probably wasn't hard for them to stay out of our way. At times near the outskirts of the forest, in more open country, we would flush a flock of blue and green or red and yellow macaws that took off screaming at being disturbed. Howler monkeys were the most common animal seen and heard; occasionally we would see spider monkeys and more rarely three-toed sloths. The latter were hard to see as their coats were partially covered with what I assume was a green mold that allowed them to blend into the vegetation.

Most nights we took the precaution of tethering the mules near our camps, leaving a Coleman lantern burning because the settlers we met told us that jaguars had been spotted throughout the area. We never saw one, but some nights we could hear their growls as they prowled around in the river bottoms or just outside the area illuminated by the lantern. We never had a problem except that, with the mules so close to camp, we could hear them at night chewing on the grass and palm fronds Manuél and Julio cut and left on the ground. Mules are noisy eaters.

At the end of the previous season, while working in the mountains east of Turbo, we met a settler who had killed a jaguar the day before. I bought the dried skin for 100 pesos. Several years later my mother had the skin prepared and for many years it hung behind the bar at our home in Rockville. I'm not sure if it is legal to keep a jaguar skin; however, someone will inherit it and decide its fate.

Both Jürgen and I were greatly interested in the animal life we encountered, especially the birds. Very little was known of the diversity and range of birds living in the area west of the Gulf of Urabá from the coast up into the Tacarcuna mountains. In 1958 Jürgen had begun collecting birds for the National Museum in Bogotá, and we continued collecting them in the Darién. At the end of each day as we rode back to camp, or around our camp site, I shot a few to add to the collection. I used a high-powered air rifle that Jürgen had brought from Germany for this very purpose. It shot a shaped pellet and was very accurate within

20 or 30 yards. Jürgen hadn't anticipated having a partner who was a good shot! We collected a wide variety of specimens from humming birds to others as large as pheasants. Most were shot with the air rifle. For larger birds like the pajuíl, a pheasant-size bird, I used the shotgun. After we had identified and skinned the large ones, the carcasses went into Sierra's frying pan.

Jürgen would measure a bird's vital statistics, determine its sex and describe its colors. Then he would skin it, shake a preservative powder on the skin, stuff it with cotton to its original size, and wrap it in a piece of newspaper. He carried a very good set of reference books that usually helped us confirm the identification of a bird unless it was a rare subspecie. We collected several hundred birds, a few quite rare, and gave them to Colombian ornithologists back in Bogotá. A number of scientific papers, including those written by Jürgen, were published based on the collections we made. The collections can be seen at the Instituto de Ciéncias Naturales, Universidad National, Bogotá. A few skins were sent to the American Museum of Natural History in New York and the Smithsonian in Washington.

Those who may think this was a cruel way to conduct science must consider that every museum with a bird collection probably has thousands of such skins filed away in drawers. That is how new species or subspecies were identified in years past. With these collections knowledge is gained of such things as range and migration routes, important to understanding how to protect tropical and migratory birds. I would agree that this was in some ways a wasteful way to learn about birds, but for many years there was no alternative. Today bird study is accomplished by catching birds live in a "mist-net" not available in 1958, noting important data, and then releasing the birds to fly another day. Thus science progresses.

A short wildlife digression: one year later, while on another mapping survey, Montez climbed a tree and took a yellow-head parrot chick from the nest. I raised him in camp and took him back to Bogotá. He became a wonderful talker. But when we returned to the States I gave him to a friend rather than put him through a long quarantine. Also that year we rescued a baby spider monkey. Initially, when he was just a few inches long, he would hold on to the hairs of my arm while I rode Nora during camp relocations. He grew rapidly in the one month I had him. When I went back to Bogotá I gave him to one of the crew. Probably he would not have been a good household pet because full-grown spider monkeys are quite large and very athletic.

In some respects the geology was less exciting than the wildlife. However, what we were able to describe and map was of great interest and importance in view of how little was known of the geology west of the Gulf. As we had hoped, by traversing the many rivers and streams we found outcrops of formations of Cretaceous age at several locations. We recorded the structural and stratigraphic geology and ended many traverses near the mountain peaks where we found formations associated with volcanic or basement rocks.

Some of the streams made deep cuts through the mountains as they cascaded and tumbled toward the Gulf, cutting overhangs in the cliff faces. Perhaps one day archaeologists and anthropologists will excavate and study these remote rock shelters that may have been used by migrants who, many thousands of years ago, traveled from Alaska to Chile. The Darién Gap also represents a gap in our knowledge of how these migrants moved south some 12,000 years ago.

In this manner we slowly worked our way north. We mapped the Rio Tanela and surrounding area but, just like King Leopold five years earlier, never came across any remains of Santa Maria de la Antigua. Of course, we weren't really looking for the city because we didn't know that we might be walking over its remains. It was only many years later, when I sent a proposal to the National Geographic Society to write an article on Balboa, that I did the research noted earlier.

While mapping west of the small town of Titumate we entered a Cuna settlement. We immediately met with the chief and requested permission to work in his reservation. We told him that we wanted to study the rock formations on the mountain peaks just west of his village, showed him the information we had recorded in our field notebooks, and told him we would leave as soon as we had finished. Understanding the purpose of our work seemed to relieve him of his suspicions, and he offered his teen-age son as a guide. We accepted and offered to pay him for his help, assuming this was an easy way for the chief to keep track of what we did and confirm we were not a threat to his people.

Leaving our mules outside the village, we walked through it trying not to seem too curious. There were so many questions I would have loved to ask, as this was our first encounter with the Cuna. My 35mm camera was dangling around my neck in clear view, but I didn't take any pictures. The village consisted of several large thatch-roofed huts on raised platforms. The area around the huts was cleared down to bare earth and was spotless. A few small children came over to look at us but the women, all colorfully dressed and with gold rings in their noses, seemed unconcerned and went about their business.

The chief's son wore a short-sleeved shirt, shorts, straw hat and sandals. Walking briskly ahead, he left a pleasant fragrance in the air, definitely not from a bottle of Gillette After Shave. I can only assume it came from some native herbs. We climbed toward the peak on a broad, clean trail. The chief's son, speaking very good Spanish, answered the few questions we asked. The tribe had its own school where he was taught Spanish. He seemed well aware of the world outside his village and had visited Cartagena, probably where he had bought his shirt. He said his village carried out extensive trade with tribes on the Panamanian side of the mountains. Reaching the top, we gazed to the west and Panamá. The rain forest lay unbroken as far as we could see. The few geological outcrops near the crest, basement intrusives of dioritic composition (a term I will leave for the inquisitive to research), were similar to those we had found in earlier traverses to the south.

Back down the mountain we said goodbye to our guide, gave him some pesos, mounted our mules and rode off, leaving behind a multitude of unasked questions about his life and that of his village, his plans for the future and how he thought his small tribe would cope with a world that might soon swallow it up. Should I have asked them? I wish I had; it would have added greatly to this tale.

Over the next weeks we continued mapping to the north and arrived, near the end of April, at Acandí. This completed the planned field work we had started in Sautatá. Acandí was the largest town on the west shore of the Gulf and just a few miles south of the border with Panamá. We returned to Sautatá, riding our mules and making small camps along the way to speed up the trip. In a few days, staying near the coast, we retraced our way quickly. At the arranged time the LCU met us at Sautatá and took us back to Turbo. We flew back to Monteria, leaving Figueroa in charge to get the crew, mules and equipment back by the overland route.

Returning to Bogotá we wrote our report that included work we had finished in 1958 on the east side of the Gulf. In June Carl Spalding sent it to Henry McAuliffe, Mobil's general manager in Colombia, and then on to Mobil headquarters in New York. During 1958 and 1959 Mobil drilled two very costly wells on the coast north of where we had conducted our field work, Cordobá-1 and La Rada-1. Both were dry holes. As a result of the drilling, the new geological interpretation that Jürgen and I had made of the area and the government enforcing more restrictive concession laws, Mobil decided to relinquish its concessions in northwest Colombia.

I spent some of the next three years mapping parts of northern Colombia, either by myself or with Jürgen and another Colpet geologist, Arno Beuther. During this time I saw many parts of Colombia that few Colombians ever visited. Perhaps of greater distinction, almost all was done on the back of a mule. If the miles covered during the many field seasons were placed end to end, I rode a mule from the Magdalena River at Mompós to the Panamanian border. And it wasn't a straight 200 miles but included many tortuous, fascinating miles in between. I wouldn't have had it any other way.

Wildcat Wellsite Geologist

During my five-plus years in Colombia I was the wellsite geologist on four wells. Three were categorized as wildcats, La Rada-1, Tenché-1, and La Heliera-1. The fourth was at the edge of a Mobil producing oil field, Cicuco, in the lower Magdalena River valley southeast of Mompós. For the latter well the only story of note is that one afternoon a Colombian worker drove a D6 caterpillar too close to the river. The bank gave way, it fell in and as far as I know is still somewhere on the bottom of the Magdalena River. The well itself was not a prolific producer.

Of the three wildcats the most interesting is La Heliera-1 drilled in Colombia's Llanos, a very large flat expanse east of the Eastern Cordillera. Wildcat wells are always important because the designation "wildcat" means that no other well has been drilled nearby. Whatever is discovered, whether or not the well produces oil or gas, adds important knowledge of the area. If it is successful, then the rush is on to secure any property nearby that is not already under lease. As a wildcat well is drilled, information on what is happening is usually kept "close to the vest" in order to keep competitors guessing and allow the drilling company to lease surrounding land at the lowest possible price if it is successful.

Mobil had vast holdings in the Llanos of Colombia that stretches for hundreds of miles from the flanks of the Eastern Cordillera to Venezuela and Brazil. It is drained by several large rivers including, from north to south, the Casanare, Meta, Guaviare, and Amazon. The Casanare is one of the headwaters of the Orinoco, the largest river in Venezuela. Because the Llanos is so flat it was not possible to do much surface geology field work; the subsurface formations could be studied only along the base of the Cordillera. To evaluate the potential of its holdings Mobil hired a contractor to run seismic lines across the concessions. Mobil geophysicists reviewed the seismic profiles and selected several places where it appeared there were subsurface structures that could hold oil or gas. Seismic profiling and analysis of such a large area is very expensive. A return on investment, as well as adherence to concession requirements, was needed; Mobil's management in Bogotá recommended a three-well program, and Mobil management back in New York City agreed to proceed.

Drilling a wildcat well anywhere in the world can be difficult because the site selected is usually well off the beaten track. Roads must be built and utilities brought in. Wildcats overseas, such as in Colombia, can be really isolated. Drilling a wildcat in the Llanos of Colombia was exceedingly difficult. The site selected for the first well was on the Casanare River in Arauca Province, about ten miles west of the small town of Rondón. Building a road to bring a drill rig to the site would have been prohibitively expensive, so a new diesel-electric rig was barged up the Orinoco from Venezuela and then to the well site, a distance of some 750 miles. A large houseboat was towed to the site to provide living quarters for all the workers. Mobil supervisors, petroleum engineers and geologists flew back and forth from Bogotá on the company's DC-3, landing at Rondón's small grass strip and then traveling by outboard motorboat to the site. We also stayed on the houseboat. I was the wellsite geologist and Harold Oliver, another Mobil geologist, relieved me a few times after drilling commenced. Based on the seismic profiles, the targeted depth of the well was 9,000 feet. However, because so little was known of the subsurface formations, we expected to drill and core below that depth, perhaps to the geologic basement.

A day's work for a wellsite geologist means examining cuttings that come to the surface from the bottom of the hole, carried up by the mud that also keeps the drill bit cool and the hole from collapsing. The cuttings are removed on a shaker and the mud pumped down again. The mud circulates rapidly, so cuttings coming to the surface and collected at the shaker are very close to the depth where the bit is making new hole at the time. The wellsite geologist continuously collects small samples; for this well I bagged samples at ten-foot intervals based on the recorded depth where the bit was cutting at the time. Then I examined the cuttings under a microscope and described the formation from which they came. With this information I then plotted a profile of the subsurface formations. The cuttings were also examined for any indication of oil staining. They were placed under an ultraviolet lamp that makes oil, if it is present, give off a blue fluorescent glow. Well cores bring complete sections to the surface rather than cutting chips, permitting more careful study and testing. This is normal procedure for wildcat wells.

Starting at the end of July 1959 drilling proceeded without any major problems, much different from my experience on La Rada-1, my first well. There, petroleum engineers had a difficult time maintaining the mud at the proper weight to keep the drill hole from caving in. A few times the drill pipe stuck deep in the hole after the hole collapsed, and a lot of time was wasted retracting the drill string and starting to drill again.

The formations first penetrated at La Heliera-1 were not tightly cemented, but we didn't encounter any problems. The first week we made almost 1,800 feet. For thousands of feet we drilled through a thick, non-marine sandstone and shale section. I could not find any microfossils in the cuttings which could have dated the formations. On and on we drilled, easily penetrating the subsurface formations. We stopped twice to case the hole to assure it would stay open. Every morning at 8 AM we radioed in the results of the previous day's drilling. Mobil had its own radio frequency allocation, but the report that included drilling status and geology was sent in code to prevent someone intercepting and following our progress.

Each time, before we cased the hole, the Schlumberger well service company ran a series of electric logs. In this procedure a special tool is lowered down the hole and electric currents are transmitted into the subsurface formations. From the logs recorded during these tests we could determine several different characteristics of the formations drilled. When the casing was run I would return to Bogotá for a few days with the well logs, returning to the well when the casing was completed and drilling was ready to restart.

Life on the houseboat was rather boring as it took only a few hours each day to stay up to date on the geology. In the off hours I water skied in the river with one of the Mobil petroleum engineers taking turns running the boat.

After a month and a half of drilling without encountering any interesting geological findings the drill bit hit a harder formation at about 8,300 feet and the drilling rate slowed. The cuttings coming to the surface were a grey-black, very dense shale interbedded with fine-grained sandstone, completely different from the previous cuttings. Now this was becoming interesting. Was this the seismic horizon the geophysicists had targeted where oil might be found? Another encouraging sign: we observed gas bubbles in the mud as it circulated to the surface. At 8,571 feet we decided to take a core to get a better understanding of the formation.

Under the supervision of Haliburton, another oil well servicing company, we recovered 50 feet of core. When brought to the surface, the rock contained in the core was very different from the cuttings I had reported from the section above. Although Oliver described it as a grey shale, I called it a slate. It was very hard with thin, tightly cemented quartz and sandstone veins interleaved in the slate. We packaged the cores up, and I took them to Bogotá for further analysis. When they arrived, after a little bumping around in transit, some of the core had split along bedding planes. Now we found preserved in the newly exposed surfaces many different kinds of trilobites and marine shells. Trilobites, a very ancient form of sea life that began living on the ocean floor 500 million years ago, became extinct in the Devonian period some 250 million years ago. They aroused my curiosity but, unfortunately, they were not studied by an expert and so the real age of the slate, to my knowledge, was never determined. I still have a few pieces containing fossils that I may have analyzed one day, as well as excellent photographs. The absence of microfossils prohibited company paleontologists from definitively dating the 8,300 feet we drilled through before hitting the harder slate and sandstone. Logically they were deposited fairly recently in geologic time, perhaps 55 million years ago when the Andes were uplifted.

At the depth where we took the cores the formation age suddenly jumped back in time 200 million years or more from that above, crossing what is called an unconformity. This is not an unusual geological occurrence, but an unconformity showing such a long time gap between formations had not been recorded previously in eastern Colombia. Reviewing the sparse literature available covering this time (papers by Hans Bürgl, Cyril Jacobs and Daniel Conley), dark shale formations containing Ordovician graptolites and trilobites were found in the Eastern Cordillera and the Macarena Mountains, the latter some 500 kilometers southwest of La Heliera. Thus, I believe that the dark slate in the core was of Ordovician age or older.

While waiting for a decision on whether or not to continue, drilling had stopped a few hundred feet below the cores. After studying the cores, logging data and our reports, Mobil's exploration department in New York decided to stop the program. The original three-well program, based on a costly seismic survey and with the next two wells scheduled at sites further to the west, was written off after drilling just one well. Mobil relinquished its Llanos concessions that stretched west to the Cordillera and north to Venezuela. In 1960 Jürgen and I were transferred to COLPET and continued field work east of Monteria where Mobil continued to hold concessions.

In 1983, 23 years after Mobil gave up its Llanos concessions, Occidental Petroleum, drilling some 80 miles to the northwest of La Heliera-1 and south of the Venezuelan border, discovered what would turn out to be the largest oil field in Colombia, Caño Limón. The producing formation is described as a thick deltaic sandstone deposited some 55 to 65 million years ago. Sound familiar? Such near misses are a common story in the wonderful world of oil exploration.

Living in Colombia

My first action on returning to Bogotá from the field was to be dewormed. Since a great portion of our work required wading in water most of the day, and I frequently had scratches or cuts on my legs and arms after encounters with thorny vines, I always returned hosting a colony of hookworms and, sometimes, a few other types of fellow travelers. If I didn't remove them, I would have had some health problems. That was easily averted; I took some pills prescribed by the company doctor and in two weeks all the companions I had picked up went to parasite heaven.

Life in Bogotá was quite enjoyable. A city of over one million, from 1958 to 1963 it was safe to travel almost everywhere. There were occasional riots; university students always seemed to be angry at someone or something. Once Pat was tear-gassed while standing in line at a movie theater, and one day while at the supermarket someone tried to kidnap our friend Phoebe Gilmore's son David. But we never felt threatened and became used to the occasional inconveniences. Besides, it made interesting afternoon cocktail conversation. We subscribed to the daily *El Tiempo* and *El Espectador* newspapers to keep up with all the goings on and to help our neighbor afford his chauffeured limousine.

When back from the field I commuted by taxi to Mobil's office downtown, a trip of about 30 minutes. The year-round climate was pleasant and, although there was a short rainy season, it seldom rained all day long. If it was a little chilly most women adopted a local costume, the ruana, a cape made of wool that came in all colors. Boys and girls could usually get by in shorts and sport shirts or cotton dresses. Men wore business suits to the office. Middle and upper class Bogotanos were very stylish, the women almost always in tight-fitting skirts and high heels. There was a large expatriate community, many living near us in the northern part of the city. There were two country clubs where we played golf and tennis, taking advantage of the Mobil memberships.

For the wives, who were mostly stay-at-home moms, there was a full social life. Husbands who worked outside the city were also invited to attend functions when we took off our boots and work clothes. There were many excellent restaurants, night clubs if we wished to go, and all the US movies, with Spanish subtitles, were shown soon after their release in the States. The restaurants served cuisine from around the world to satisfy the many expatriates who lived in Bogotá.

Most Bogotanos, except those working for US companies, observed "siesta" and afterwards would go to the many coffee shops serving delicious high-calorie confections to tide them over until dinner before going back to work late in the afternoon. If we went to a Colombian friend's home for dinner we would expect to be served sometime after 9 PM. Our dinners at home were eaten at a conventional time. US companies worked the normal eight to five with a one-hour lunch. Those of us who spent long periods in the field were allowed a little leeway, and our lunches sometimes lasted longer.

Our home was comfortable; it had all the modern conveniences and then some. We hired a full-time maid, Alícia, who had been recommended by an American family returning home. She was a young indian woman of 18, short, stocky, with bronze skin tones, high cheek bones, dark and narrow slanted eyes, and coal-black hair always neatly tied in a pony tail. She had her own bedroom and bathroom off the kitchen. She helped in all household duties, including some cooking (she made perfect steamed rice lightly flavored with a few onion slices), and was always available as a baby sitter.

Alícia was an excellent companion for Tommy, at first pushing him around the neighborhood in a stroller and later walking with him to a nearby park where all the maids gathered to show off their charges. She didn't speak English, so she taught him Spanish; as he grew older he became completely bilingual with a perfect accent. She had one day off each month and was paid $50 a month plus uniforms and food. Our Colombian neighbors thought that we paid her too much.

We also had a part-time maid who came two days a week with her young daughter to do the heavy cleaning, washing floors and windows, and the laundry. Colombian households had a well understood hierarchy; the live-in maids weren't expected to do this type of work. Twice each night a watchman came by and put a slip of paper with the time noted under our door to show he had been there. I believe I paid about one dollar a week for the service. Perhaps this was a form of blackmail to avoid problems; all our neighbors used the same service. But we never had an attempted robbery or had anything stolen.

In addition to Mobil families that we knew and socialized with, we became close friends with a family that lived a few doors away, Bill and Phoebe Gilmore. Bill was the Ingersoll Rand representative in Colombia. Their first son, David, born in Bogotá, was about a year younger than Tommy. A second son, Tom, was born two years after we arrived. In 1960, at the request of some of our Colombian friends, Bill and I started a Boy Scout troop, Troop 1. I'm not sure if that meant we had the only official troop in Colombia, but it was the only one at that time in Bogotá. I started as the scoutmaster, but I was often away; Bill "volunteered" and was scoutmaster for the next two years until he was transferred to England.

We had both Colombian and expat boys in the troop. One of the wealthy Colombians donated a campground about three hours from Bogotá, part way down off the highlands. At this lower elevation it was warmer and made an ideal camp site, as it was surrounded by semi-tropical forest. We would take the scouts and a few dads camping whenever I was in town. At the end of the first year the troop selected the two outstanding scouts and I took them for a week, via the company plane, to stay where I was working. We put up a separate tent for them and they accompanied us, riding mules, while I did the mapping. Whether they wanted to or not, they received a short course in geology during their stay; I wonder if those two boys remember that experience.

In 1961 I helped organize a fast-pitch softball team composed of players from Intercol and COLPET. There was an eight-team league in Bogotá, and we were the only team made up mostly of "gringos." I was the pitcher, and in 1962 we won the championship. In 1963, my last year in Colombia, we were at the top of the league when I left.

In 1961 Jürgen married Maria Kluge, who had come to Bogotá from Germany to teach at a German school. There was a large German community in Bogotá, many having lived in Colombia for generations. We knew several, and one, Hans Bürgl, was a noted paleontologist who often provided professional services to Mobil and COLPET. The wedding reception was held at a lovely restaurant in downtown Bogotá, formerly a bishop's palace, the Gran Vatél. It was a grand affair, attended by Mobil and COLPET families and the Haffers' German friends. Maria was a beautiful bride, and we were all impressed with Jürgen's good fortune in finding a wife so far from Germany. After the reception they left on a three-month round-the-world honeymoon (Jürgen had been saving up his vacations). After a short assignment in Germany they returned to Colombia in March 1963, and Jürgen rejoined COLPET. Both their children, Christian and Amelie, were born in Colombia. When I left Colombia I gave Jürgen my Zenith radio in case he should find himself living alone again in a remote camp.

On two occasions I took Tommy with me on the company plane to Coveñas. My new partner Arno Beuther and I had finished some work in the San Jacinto Mountains east of Coveñas and were staying in one of the unoccupied small houses at the pipeline terminal rather than a camp site. So it was easy and convenient for Tommy to be with me. COLPET had been running seismic lines farther to the east, and we took advantage of the helicopter supporting the work to get an aerial view of one of the structures we had mapped. On the other trip I was working in the immediate vicinity alone with Figueroa and the crew. We drove to outcrops during the day and ate dinner with the terminal manager's family. We walked out on the long floodlit pier at night and watched the fish that were attracted by the lights.

For family vacations we went to Cartagena and stayed at the Hotel El Caribe, the only first-class hotel in Cartagena at the time, outside of the city on the beach at Bocagrande. It was a nice change for Pat to get down to the "hot country" with a pool, sightseeing at the Spanish fortress, and some shopping in the old walled city. Vacationing at the coast was a pleasure for all who lived in Bogotá as we noticed that upon arriving at sea level breathing heavier air gave us added energy.

Every year the Colombian Society of Petroleum Geologists and Geophysicists scheduled a field trip to places of interest. As these locations were usually well outside Bogotá, they required overnight stays. At the beginning of 1961 John Closs, my new boss at Mobil, and I volunteered to do the advance reconnaissance for a trip to the famous emerald mines at Muzo some 90 miles north of Bogotá with several side trips to important geological sites. We also arranged hotel accommodations and meals. For the trip we proposed, the first day would include several lecture stops before arriving for dinner at the Hotel Aguas Blancas in Puente Nationál.

On the second day the tour would continue to the mines and return to Bogotá that evening. The mines supplied emeralds of the highest quality, some worn as jewelry by the pre-Colombian Chibcha. After the Spanish subdued the local tribes they continued to mine for emeralds, using the indians as forced labor. The mines were of interest to geologists not only for the emeralds but because of their unique occurrence in dark shales of Lower Cretaceous age.

Early on the morning of March 25 the participants left Bogotá by bus with guidebooks supplied by Texaco under their arms. It was a successful trip, with the lectures provided by several company geologists giving the attendees an excellent overview of the complex geology of Colombia's Eastern Cordillera.

1961 was an auspicious year in many respects; however, the big news in our family was the birth of our second son, Bruce. He was born on Flag Day, June 14, in a Bogotá hospital. We were concerned about how Pat's delivery would be handled as her Spanish was limited to directing the maids and talking to shopkeepers. As our friend Phoebe Gilmore's two sons were born in Colombia, it was comforting that she had had a good experience. Our concerns soon evaporated as all the Colombian doctors spoke very good English. Pat's delivery went without problems and, since it was the end of the field season, I was in Bogotá to take her to the hospital and stay until Bruce was born. He can now have dual citizenship if he wishes. I hope not!

Leaving Colombia

At the beginning of 1962 as I prepared for a new field season I could foresee that my time in Colombia might soon end as little by little Mobil and COLPET gave up concessions. With the reduction in exploration work some of the staff had been transferred to other countries where exploration was expanding. Our former paleontologist was now working in Libya. I had no way of knowing where I might be assigned if Mobil's Colombian operations were closed down.

John Glenn was at Cape Canaveral, Florida, in February 1962 preparing to become the first American to orbit the Earth and I was working with Arno Beuther west of the San Jacinto Mountains in northern Colombia. Even in this remote area we picked up Armed Forces Radio and Voice of America on my Zenith radio. We were closely following the space launches of the newly formed NASA, agonizing along with everyone back in the States over the failures and delays as we tried to catch up with the Soviet Union's aggressive space program.

After each of several launch delays NASA projected a new lift-off time and, based on those projections, we attempted to complete our daily fieldwork and get back to camp in time to hear the launch broadcast. When you were far from home and your immediate world was bounded by a small rain forest camp and how far you could ride each day on the back of a mule, it was easy to become absorbed in the drama at Cape Canaveral. During one of the delays before Glenn's launch, the announcer filled in some air time with interviews. One was with a gentleman from NASA's Public Affairs Office who described the Apollo Project, what little was known of it at the time. He mentioned that to accomplish Moon landings NASA would be hiring geologists to assist in mission planning. He provided an address where interested applicants could apply. My curiosity was piqued. I copied the address, pulled out the rusty typewriter we used to write our monthly reports, and composed a letter to NASA. I explained that I was not only a geologist but an ex-Navy jet pilot and felt that I would fit right in with NASA and the astronauts that were being selected.

Eventually John Glenn was launched. In Bogotá on my next rest days I mailed my letter off, convinced that NASA could not turn down such outstanding qualifications. In my naivete I even thought I might have a chance to become an astronaut. What better combination of experience for someone going to the Moon, I reasoned, than a geologist-jet pilot, especially one accustomed to working in strange places under difficult circumstances such as coexisting with army ants, vampire bats and jaguars? With some modesty my letter included these qualifications. It was several months before NASA replied. It was a very polite letter thanking me for my interest and enclosed application forms to submit to the Goddard Space Flight Center in Greenbelt, Maryland. I filled out the forms and mailed them back. Then began a wait, now with some anticipation since NASA's reply had been so encouraging.

At the start of the rainy season I was back in Bogotá when another envelope arrived telling me that I had qualified as a GS-13 Aerospace Technologist, Lunar and Planetary Studies, and that my application was being circulated in NASA to determine if a position was available. Aerospace Technologist? I wasn't sure what that meant, but it sounded impressive. I had visions of doing exciting things at this new agency and helping to send men to the Moon. Then began what turned out to be a longer wait. In December I received another letter saying my application had been circulated but no positions were open; they would keep it on record in case one turned up. Rejection! That didn't fit in with my plans, and I resolved to pursue my quest the next time I was in the States.

On my next leave in June, 1963, I went to Washington to talk directly to someone at NASA. I bought an aerospace trade journal with the latest NASA organization, complete with names. In it I found an office at NASA Headquarters where my background and interests might best fit: Lunar and Planetary Programs in the Office of Space Science (OSS), headed by Dr. Urner Liddell. From my family's home in New Jersey I drove to Washington without an appointment and went to Dr. Liddell's office. With great good fortune, as I learned later, Dr. Liddell was traveling that day but his deputy, Dr. Richard Allenby, was in. The good fortune was that Liddell turned out to be a rather formal bureaucrat and probably would not have given me an interview without an appointment. Dick Allenby was just the opposite and agreed to see me.

After a brief introduction I learned that Dick was an old oil field hand (geophysicist) who had worked in Colombia just a few years earlier and that we had several mutual friends. We hit it off at once, marking the beginning of a long professional and personal relationship. Dick liked my background but had no openings. He set up a meeting with Navy Captain Lee Scherer, who had just been hired to manage the Lunar Orbiter Program (satellites that would orbit the Moon to photograph potential Apollo landing sites). Scherer was not hiring at the time either but thought he knew someone in the Office of Manned Space Flight who needed a person with my experience. I was beginning to question my timing. Lots was going on at NASA with new offices being set up all over town but perhaps no one with the need to hire, just as the last NASA letter stated. Lee, who would become my boss six years later, set up a meeting with another NASA military detailee, Major Thomas C.

Evans, US Army Corps of Engineers.

With Lee Scherer's introduction getting me in the door, I introduced myself. During our conversation I learned that Evans had been a project officer during the building of Camp Century in Greenland, 800 miles from the North Pole, the first successful adaptation of nuclear power for a military ground base. His background was ideal for the job NASA had assigned him, the design of a future lunar base. Camp Century was about as close to a lunar base as one could come on Earth. Tom was an impressive officer, later to become a congressman from Iowa, and he spent the next hour or so telling me about his office's responsibilities to plan a lunar program that would follow a successful Apollo Program. He was enthusiastic and brimming with ideas, the kind of leader that everyone looks forward to working for. Best of all, he thought I could help the team he was putting together. Since it was getting late in the day, Tom asked me to return the next morning to talk to his deputy, Captain Edward P. Andrews, US Army, to determine how we would proceed.

My discussions with Ed the next day went well and, since I had already received a civil service job rating, Ed proposed to start the paper work. Two days in Washington and I was offered a job as a "lunar aerospace technologist" at what I considered to be the most exciting place in town, NASA! I would receive the princely sum of $11,150 per year, less than my Mobil salary, but I felt I couldn't pass up the opportunity. Ed told me he would call me in Colombia when everything was final and told me to start planning on moving my family to Washington.

Returning to Colombia, I began to close out my work. My coworkers all thought I was crazy to take on such a job. Most thought that trying to get a man to the Moon was quixotic at best and probably couldn't be accomplished. A job that entailed planning for what would be done after we landed on the Moon was real science fiction. In turn, I thought they failed to appreciate the beginning of a real adventure. In August I received the phone call I was waiting for. Ed said all was in order and they were waiting for me to arrive. Smugly I filed away my NASA correspondence, including the "not now but possibly later" letter. At the end of August I left Colombia with my family on a Pan Am Boeing 707 to begin a new calling, one that never lost its thrill and satisfaction over the next ten eventful years.

Scene from my thesis area

Thesis camp visitor, Pat

Green Dragon - not the real one and in much better condition

F9F-6 at NAS Buckley Field, Denver

TV-2 trainer (above), Map of Northwest Colombia (below)

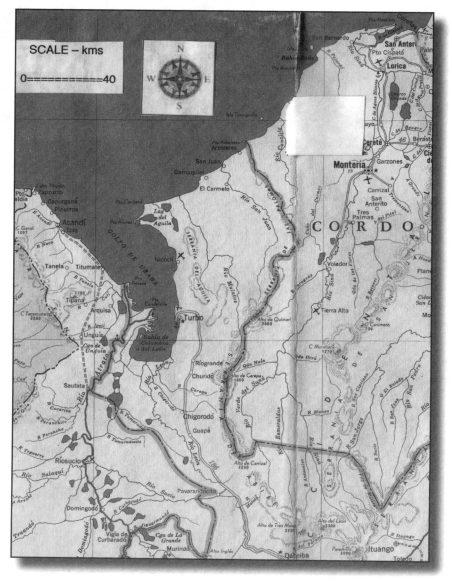

Camp site - Quebrada Resbalosa (above),
Quebrada Resbalosa - home to the Morpho iris butterflies (below)

Camp site in Cansona area

Our rain forest home and office, Ñecas

At Boca la Balsa - Figueroa, Elias Lopez, I on my trusty mount Nora, Julio Caro

Our cook, Sierra

On the Rio Sinú with Pedro Valencia and Julio Caro

Tacarcuna Mountains looking to northwest from the Atrato River

Mot-mot collected in the Darien. The faint mustache died too

Monteria boulevard along the Sinú river

Motorless ferry at Tierra Alta

Monteria with Arno Beuther, Tommy and helicopter pilot, 1962

Casanare River at La Heliera wellsite

Alicia in kitchen feeding Tommy and Bruce

Tommy and Bruce in the shower

Reading to Tommy and Bruce in Bogotá home

Bogotá from Monserrate

Hotel Caribe, Cartagena

Chapter VII
Inside the Beltway - 1963 to 1983

Today, many who work in the Washington D.C. area labor inside the I-495 Beltway that circles our nation's capital. As a result the term "inside the beltway" is applied, often negatively, to those Washington bureaucrats who believe themselves to be very important. They usually work for the government or related institutions. When I began working at NASA the Beltway didn't exist. I joined a large number of people just trying to do our jobs. But soon the Beltway was completed and I worked within its borders for many years.

National Aeronautics and Space Administration 1963 to 1973

Back in the States, we left Tommy and Bruce with Pat's parents in Connecticut and drove to Washington to house hunt. I knew that most of the staff in my new office lived in Virginia, but when I considered commuting, schools, housing prices and other concerns it seemed that Maryland might be the best choice. We looked in Greenbelt, Maryland, a new planned community in Prince George County, and north of DC in Montgomery County. The farther north we went the more affordable home prices became.

We found a nice four-level, three-bedroom home in Rockville for $25,000; mortgage assistance under the G.I. Bill required only ten percent down. It had been the home of a pastor provided by his church. One room on the ground floor was finished as an office. The elementary school was only two blocks away, and Montgomery County reputedly had the best schools in the DC area. It seemed to be an ideal location to bring up two boys. Tommy had attended kindergarten at a school for expat students our last year in Bogotá, so he was ready for first grade.

New homes were springing up like weeds in the neighborhood. Our nextdoor neighbor, Ernie Neal, had moved in a year earlier and was employed at the Goddard Space Flight Center. Down the street a NASA Headquarters employee, Bob Rollins, moved into a just completed house with his four children. It was definitely a middle-class neighborhood of young professionals and stay-at-home moms.

Settled in, I began my career at NASA as a GS-13 in the Office of Manned Space Flight, Manned Lunar Missions Studies. NASA offices were spread all over town. Some, including mine, were awaiting completion of a government building at 600 Independence Avenue, SW. It would put several of the scattered offices under one roof, but in September 1963 our office was at 1815 H Street, NW, just a few blocks from the White House. We shared the building with other organizations and other NASA offices, including those for Manned Planetary Missions, Systems Engineering, Launch Vehicle Studies and other advanced studies.

I was assigned to an office with another recent hire, Tom Albert, a mechanical and nuclear engineer who was determining how to modify the planned Apollo systems to enable longer stay times and lunar base missions. Since my work at Mobil was primarily writing reports based on what we had accomplished in the field, Tom really impressed me. He spent hours on the phone talking to NASA and private company engineers, taking a few notes and going on to his next call, all the while speaking a language I didn't understand in which every third word seemed to be an acronym, NASA-speak. It was confusing and annoying at first, but I soon started to catch on and quickly moved to the next level, inventing acronyms for my own programs. This new skill brought a real sense of power and control. I am convinced that NASA could not have functioned without these shortcuts. It became an unspecified requirement that new programs come up with catchy acronyms that would appeal to the ears and eyes of management, Congress and the media. During meetings most of the discussion was based on acronyms spoken like real words.

Our office consisted of seven engineers with diverse backgrounds. Tom Evans and Ed Andrews shared one secretary and the rest of us shared the other. Two or three engineers occupied each office space; new arrivals were assigned interior offices while those with windows were reserved for the more senior staff. These were fairly spartan accommodations, but there were few complaints since we knew we would soon be moving to a new building. There was one empty desk in the office I shared with Tom; I was told it had been occupied part-time by Eugene Shoemaker, who had been detailed from the US Geological Survey (USGS). He was on his way to Flagstaff, Arizona, to start a new USGS office. Thus I missed meeting Gene by a few days, but in a short time our paths would cross and we would work closely together until the end of Apollo.

My first days at NASA involved the usual getting acquainted. The Navy had also been a government bureaucracy, but NASA functioned quite differently. Owing in part to Tom Evans' style and the fact that NASA was a new agency with an unprecedented mission, multitudinous rules and procedures had not yet been instituted and the staff was given great freedom of action. During my six years with Mobil I had usually set my own daily schedule, so this was an ideal situation for me. With Ed Andrews' guidance I immediately began to define my role and make the contacts at NASA and in the scientific community that would make my job easier.

Science Planning for Apollo and Post-Apollo

I soon learned that Shoemaker had come to NASA to help bridge the wide gap between its science side (OSS), where I had made my first contact, and the Office of Manned Space Flight (OMSF). Major differences had surfaced between the two offices resulting in a power struggle over how to apportion NASA's budget. The debate about specific science experiments to be carried on Apollo missions still lay ahead. OMSF was already receiving the major portion of NASA's budget, and OSS staff, as well as scientists outside NASA who looked to OSS to fund their pet projects, were constantly fighting to persuade top management to change funding priorities. These efforts were led by such luminaries as James Van Allen, who had made one of the first space-based science discoveries of the radiation belts surrounding Earth, which were later named after him.

The complaints were reinforced by the National Academy of Sciences and its Space Science Board that provided advice to Homer Newell, the OSS Administrator. I was told that Shoemaker, during his brief stay, had begun to reduce some of the distrust that had developed but had only scratched the surface. Apparently it would take more than his talents to resolve the differences. Unfortunately, almost 50 years later this budget power struggle continues, inside and outside NASA.

With Tom Evans' blessing I was given an unofficial second hat to work with both OSS and OMSF on matters dealing with lunar exploration. Upon Shoemaker's departure another USGS detailee, Verne C. Fryklund, who had been working on Newell's staff, took his place. Verne was definitely from the old school, gruff, with a bushy mustache and a half-smoked but unlit cigar always in his mouth. He looked quite professorial in a tweed jacket with leather elbow patches. He was given the title of Director, Manned Space Sciences Division (Acting), Office of Space Science. In Washington's bureaucratic culture having the word "acting" attached to a title usually meant the position was temporary until the person management wanted for the position was found and installed.

Fryklund's primary duty was the same as Shoemaker's, to be the go-between for the Office of Space Science and the Office of Manned Space Flight. During his shuttle diplomacy Fryklund tried to represent the interests of the science community to NASA's manned space side, directed at the time by Joe Shea and viewed as being unfriendly to science. Fryklund became my unofficial second boss. His staff was appropriately small, consisting of a few scientists detailed to his office such as geologist Paul Lowman from the Goddard Space Flight Center (GSFC) and others. Thus he was happy to have me, if only part-time.

Fryklund, an experienced bureaucrat, approached his new job cautiously. The complicated politics were evident to someone with his background, and he was fully aware of the gulf that existed between the two organizations. Until this time nothing had been officially decided about what science would be carried out on the Apollo missions, and this became his first priority. Shuttling back and forth between high level meetings at OSS and OMSF, he relied on a draft report on scientific aspects of the Apollo program, commonly referred to as the "Sonnett Report" after its chairman, Dr. Charles P. Sonnett of the NASA Ames Research Center. It served as his guide and point of departure for his arguments on what needed to be done for Apollo science.

The draft report contained wide-ranging recommendations that included geology mapping, sampling and geophysical activities and experiments as well as surface physics, atmospheric measurements, and particles and fields experiments. Guidelines had been provided to Sonnett's committee decreeing that science experiments would be limited to what could be done using a payload of 100 to 200 pounds. However, the recommendations the committee made could not be carried out if limited to that weight. Geology traverses up to 50 miles from the landing site were described. Sample collection, including drill or punch core samples, was also detailed and potential landing sites suggested The report even went so far as to describe what type of astronaut should be on the flights and the recruiting criteria for finding them.

At the time the Apollo program began the fundamental lunar questions being debated by planetary scientists could be quickly summarized: How old is the Moon? How was it formed? What is its composition and what caused the formation of the thousands of craters, large and small, that can be seen? Finding the answers was the driving force behind the experiments recommended in the Sonnett Report and justified the need to reserve a large science payload to carry the experiments and equipment the astronauts would deploy.

The answers to some of the questions, such as the age of the Moon, would come in large part from returned samples which might also tell us about the origin of the craters. It was expected that Apollo missions landing at interesting points on the Moon, deploying the various recommended geophysical experiments and conducting geologic traverses, would answer all the above questions. It was anticipated that we would also better understand the Earth, especially its early history that is difficult to study because of the constant changes that have taken place on its surface over the past three or four billion years.

Based on the analyses we had done of the Sonnett recommendations, in early October 1963 Fryklund sent a memo to Robert R. Gilruth, Director of the Manned Space Center (MSC) in Houston. Importantly, the memo was supported by both of Frykland's bosses, Homer Newell and Joe Shea. It contained the first official scientific guidelines for the Apollo Project. As is the nature of guidelines, they established a broad framework within which to plan but provided no specifics on critical items such as how long the astronauts would be on the Moon, how far they could travel, or how much payload weight would be allocated for science. The 100 to 200-pound numbers given to Sonnett were not included; specific numbers would come later.

Fryklund's guidelines, eight in total, listed three functional scientific activities in decreasing order of importance: comprehensive observation of lunar phenomena, collection of representative samples, and emplacement of monitoring equipment. Meanwhile, back in Washington we began to flesh out the guidelines by translating the Sonnett recommendations to some hard numbers. From all the information we could collect it was evident that the range of measurements and activities recommended would require a science payload far exceeding 200 pounds.

One month before Fryklund issued his guidelines, unbeknownst to NASA Headquarters MSC had jumped the gun and hired Texas Instruments to define experiments and measurements to be made on the lunar surface based on MSC guidelines. The report, when it was eventually issued in 1964, was dismissed as amateurish by Headquarters and members of the scientific community who had begun to focus on Apollo science. This difference of perspective signaled the beginning of a clash between Headquarters and the small MSC science staff as to who would be responsible for defining Apollo science.

In late October 1963, returning from one of the frequent meetings, Fryklund rushed into the office we shared and announced, "They have just agreed; we have 250 pounds for science!" "They" were Manned Space Flight senior management. To me as a latecomer to what had been a major struggle, 250 pounds seemed like a minor victory; 1,000 pounds would have been better. But it really was a victory and we quickly moved to capitalize on the opportunity to define a complete payload based on that "official" number.

While all this was going on, major changes were taking place at NASA

Headquarters. New organizations were being created almost weekly, and the staff was expanding rapidly. In September George Mueller was hired from Space Technology Laboratories to lead Manned Space Flight. Toward the end of the year our office was merged with several others, and the new organization was given the title Advanced Manned Missions Programs. Edward Z. "EZ" Gray was hired from the Boeing Company to be our leader. We moved to our new offices at 600 Independence Avenue where I was assigned to a two-man office on the first floor. During 1963, the year of my arrival, Headquarters almost doubled in size. In January 1964 Major General Samuel C. Phillips was detailed from the Air Force Ballistic Systems Division to become Mueller's Deputy Director, Apollo Program, and soon had his title upgraded to Director.

In the wave of reorganization Fryklund's tenure as acting director was short-lived. Homer Newell, in agreement with Mueller, formally established the Office of Manned Space Science, reporting to both their offices. Willis Foster was brought in from DOD as the new full-time director, and Fryklund became Foster's Chief of Lunar and Planetary Sciences. Foster's office, starting with Fryklund's original staff of eight, grew rapidly in 1964. To my surprise my thesis advisor at the Colorado School of Mines, Peter Badgley, was hired to develop a space-based Earth observation program. Dick Allenby was transferred from OSS to become Foster's deputy. Will Foster now became my unofficial second boss, and I continued to work on developing the science payloads for both Apollo and post-Apollo. All this detail sets the stage in order to understand the conflicts encountered in the next years.

Contrary to what one might think, detailed planning and analyses for post-Apollo science missions started before such planning for the Apollo missions themselves. The program under which we conducted post-Apollo planning in Tom Evans' office was given the name Apollo Logistics Support Systems (ALSS). The name implied a program to come after the Apollo missions and capitalize on the Apollo hardware then under design and manufacture. It was given other names in later years as NASA management attempted to get a commitment to continue lunar missions after the initial Apollo landings.

By late 1963, except for the Sonnett Report, very little had been done within NASA to fill the void in Apollo science planning; many would have denied a void existed. The Apollo program had only one objective, to land men on the Moon and return them safely. Although not defined, management probably considered that the astronauts would be permitted to take a few pictures with a camera yet to be selected and pick up a few rocks. But tools to accomplish this were not under development nor were the special boxes needed to store such samples for the return trip, a critical aspect of any sample return. Only a few forward-looking scientists were beginning to think about these matters, and no one was receiving NASA funds to develop this equipment. Post-Apollo planning was an entirely different matter. Our office was already spending NASA funds to determine what science and operations should be carried out on the Moon after the initial landings.

In the fall of 1963, less than six years before the first Apollo Moon landing took place, time lines had not been developed to tell us how long the astronauts would, or could, stay on the lunar surface. We understood it would be very difficult to get a larger allocation than the 250 pounds until all the Apollo systems had been tested and flown and their performance evaluated. In spite of the many uncertainties and lack of firm numbers, we took as a given that the Apollo landings would be successful and that all the Apollo systems would function as advertised.

Our job in Tom Evans' office was not to question any of the Apollo assumptions. We were charged with taking the basic Apollo hardware and expanding its capabilities far beyond the original intent. For example, we had contractors studying how to upgrade the Lunar Module (LM) to carry a much heavier payload to the Moon's surface, to extend the Command and Service Module (CSM) stay-time in lunar orbit, and to increase the potential landing area accessible to the LM (restricted for the first Apollo landings to the Moon's near side, central longitude and equatorial region) so that we could explore what appeared to be critical geological sites far from the Apollo landing zone. And would it be possible to land a modified, automated LM, turning it into a cargo carrier to bring large scientific and logistics payloads to the Moon?

By the time I arrived some of the preliminary studies on expanding the versatility of the Apollo hardware were already completed. They had examined the Apollo spacecraft then under development and documented its inherent flexibility. With what we claimed would be minor modifications it would be possible to land the LM at sites with no crew on board. Such an LM could then be a cargo ship carrying as much as 7,000 pounds to the lunar surface by trading off ascent fuel and other equipment not needed for a one-way, unmanned trip. An LM with this payload capacity could carry living quarters, large science payloads, or other types of equipment depending on the mission. It appeared feasible that a crew of two astronauts, arriving in another modified LM, could land close to one or more logistics LMs landed earlier unmanned and spend as much as two weeks on the Moon. During the two weeks they would either transfer to the earlier landed LM or utilize other types of payloads that had preceded them.

Similar studies of the CSM showed that it could be extended in lunar orbit to support a two-week stay. In addition, it would be possible to carry remote sensing payloads in one of the CSM's bays to map the lunar surface in various parts of the electromagnetic spectrum, an application that had begun to receive more and more attention and support.

The missing ingredient in all this planning was the reason for staying longer on the lunar surface and the need for bigger payloads. It became my responsibility to provide the reasons. At the end of July 1963, as one of his last actions at Headquarters, Gene Shoemaker had sent a letter to Wernher von Braun, the Marshall Space Flight Center Director, requesting that his staff suggest scientific activities to be undertaken on post-Apollo missions. Fryklund, as Shoemaker's successor, continued to follow this inquiry and I, in turn, inherited it when I joined his staff.

I found that Paul Lowman in Fryklund's office shared my enthusiasm to study and explore the Moon. It seemed quite natural to ask Paul to join me on an informal basis to work on some of the projects I had begun. Paul had already made a name for himself by convincing the Mercury astronauts to use Hasselblad cameras on their flights to photograph the Earth's surface. This was no mean accomplishment, as these former test pilots were much more interested in flying and monitoring spacecraft systems than in taking pictures of places Paul asked them to photograph.

As it turned out, taking photos eventually proved popular among most of the astronauts, especially when their photos were published extensively in newspapers and magazines. *Life Magazine* had an exclusive agreement with the astronauts to publish first-person accounts of the missions, and in the articles that followed each Mercury flight a few beautiful full-color photographs of the Earth appeared. As a result of this success, Paul continued to coach the upcoming Gemini astronauts on taking similar photographs.

Paul had originally accepted his temporary Headquarters assignment in order to work with Gene Shoemaker. With Gene's departure, the reorganization of Fryklund's office and the arrival of Will Foster, the timing was right for our collaboration. One of the attractive aspects of working at NASA in those early days was the great freedom given to staff members to attack whatever problem they uncovered without a lot of bureaucratic red tape and worry about whose turf was involved. Thus Paul and I began a long professional association and a friendship that endured until his death in 2011.

Most of my office colleagues had degrees in electrical, aeronautical or mechanical engineering and little training in the Earth sciences, as was the case for NASA senior management. Paul and I decided that the best way to convince our bosses that there would be exciting and important investigations for the astronauts to undertake on the Moon, requiring many days and a wide variety of equipment, was to illustrate these tasks with terrestrial analogies and describe the type of field work and experiments that were required on Earth to unravel its history.

Based on the Sonnett Report and our own knowledge and experience, the first thing we did was go to the Smithsonian Museum of Natural History to visit their rock collection. Our purpose was to select and borrow a variety of rock types that illustrated both the Earth's geological diversity and the complex geological and geophysical problems that we believed would be encountered on the Moon.

With our rock collection as visible evidence of how the planetary body Earth evolved we developed a rudimentary "show and tell," a short course in terrestrial geology and geophysics for NASA decision makers, and extrapolated this lesson to the Moon. We hoped our rock collection and other material such as maps, photos and cross sections would stimulate their interest and demonstrate that what we were proposing was real and important. We selected igneous, metamorphic and sedimentary rock types. We later added a few specimens collected at Meteor Crater, Arizona, that showed how rock formations at the impact point were modified from their original condition after being struck by a meteorite.

In 1964 so little was known of the physical characteristics of the lunar surface that we felt free to use almost any type of rock to tell our story. We put together a half-hour lecture in which we passed our rock collection to the audience at appropriate times to make a point and, hopefully, elicit questions. We started with my office, honed the presentation, and later lectured to senior staff. Tom Evans and EZ Gray were impressed with the story we put together. We were ready to take our show on the road and combine it with recent post-Apollo study results that indicated missions were possible that would allow the astronauts to stay on the Moon for two weeks utilizing sophisticated science payloads.

In December Tom Evans and I briefed the President's Science Advisory Committee and Office of Science and Technology. I answered a few lunar science questions and thought the briefing went well. Evans wasn't so sure. Tom's evaluation proved correct. Although President Johnson announced that we were not only going to land on the Moon but explore it, his administration never provided the resources to continue lunar exploration after the Apollo missions. In fact, after fiscal year 1965 NASA budgets declined.

Despite Tom's pessimism, and not knowing what the future held, we charged ahead assuming that NASA would continue lunar missions after the Apollo program was successful. Senior NASA management, in the person of George Mueller and his immediate staff, were our next target audience for our geology lesson. His schedule was usually so full that we had learned to schedule important meetings on Saturday or Sunday when interruptions were minimal and the atmosphere more relaxed. NASA Manned Space Flight under Mueller became a seven-day-a-week job and lights burned late as most offices struggled to keep up with the rapidly evolving program. Our briefing was held on a Saturday.

It was carried out in a fairly informal atmosphere with fewer than normal attendees at his meetings. We were able to make our case for longer stay times and larger payloads. Our approach and props succeeded beyond our expectations; eventually EZ Gray used them for his own briefings. Mueller too began to lobby for post-Apollo missions. Over the next two years, as more information on the Moon's characteristics became available through new studies and data returned from the unmanned missions, we improved our story and eventually won the struggle to convince NASA management that conducting science on the Moon was important and doable.

To better understand the constraints astronauts might face on the lunar surface, at the end of 1963 I began my first contractor study with Martin Marietta. Martin had lost the competition with Grumman to build the lunar module. During that competition Martin had built a full-scale mockup of their concept of a LM. Not surprisingly, since both were bidding to the same specifications, the Martin concept was very similar to the Grumman design. The Martin mockup now sat in a high bay building at the Martin plant at Middle River, Baltimore. Disappointed in the loss, and learning of our activities, a Martin manager came to my office one day to see if we had any interest in using this equipment. I had just completed a parametric analysis of contingency experiments for Apollo and saw the opportunity to determine, in a preliminary fashion, what difficulties an astronaut might have in making observations from the LM on the lunar surface before setting foot outside. I could foresee that, even after a successful landing, something might happen to keep the astronauts from stepping out on the lunar surface.

Because Martin had the only available look-alike version of an LM I was able to justify a sole-source contract. The contract required Martin to construct a simulated lunar surface surrounding the LM mockup on the floor of the high bay building as best it could within our funding limitations. To achieve the proper effect tons of cinders, sand and other material were poured on the floor. In addition we brought in different types of rocks and scattered them in the loose, finer-grained material, including some of the rocks we had borrowed from the Smithsonian. To simulate lighting conditions the astronauts might encounter on the Moon, we were able to illuminate the surface with low to very intense light and to vary the light angle to duplicate the changing sun angles that might confront the astronauts depending upon when they landed.

Since this simulation was to have human factors as well as geological factors, the contract was managed by the Martin human factors department. Test subjects, Martin employee "astronauts," were given some rudimentary geological training by Paul Lowman and me. We concentrated on how to make visual observations, provide verbal descriptions using geological terms, and take photographs from the LM windows to show the nature of the surface. The test subjects volunteered to spend three or four days isolated in the LM mockup, eating and sleeping in the confined space and only able to communicate with the test engineers by radio.

The living conditions inside the Martin mockup, although somewhat uncomfortable, were considerably better than those eventually faced by Armstrong and Aldrin five years later during the first lunar landing and by astronauts in later missions. Armstrong and Aldrin, for example, were able to get little rest during their 20-hour stay. When they attempted to sleep, Armstrong was forced to rest on top of the ascent stage rocket motor casing and Aldrin curled up in a confined space on the LM's floor. Neither slept very soundly, Armstrong perhaps not at all. We were easier on our test subjects; we gutted the interior of the mockup and each test "astronaut" had enough space to sleep on a thin air mattress.

The first problem to overcome was how to photograph and describe the scene outside the LM as it had only two small windows, both facing in about the same direction. With this limited view, less than half the lunar surface would be visible if the astronauts could not get out. The LM also had an overhead hatch to allow entry from the CSM while in lunar orbit. In that hatch was a small window designed to permit the astronauts to make star field sightings, if needed, to update the LM's guidance and navigation system. But, with the LM on the lunar surface, this window would face only the black sky above the Moon. For one test the overhead hatch on the Martin mockup was opened and one of the test subjects stood in it to make observations. That worked quite well, and we were confident that the real astronauts would be able to provide a good description of the landing site supplemented by panoramic photographs if this were permitted. On the Apollo-15 mission Dave Scott was the only astronaut to do this, not because he wasn't allowed to leave the LM but to plan his EVAs. But what if astronauts were unable or not permitted to open the hatch?

The LM would be equipped with a small telescope that could be operated from inside to assist in the star sightings. Perhaps we could adapt the telescope to operate more like a periscope, allowing the astronauts to scan the surface in all directions. Paul and I traveled to Boston to visit the Instrumentation Lab at MIT that had the NASA contract to design the guidance and navigation system for the CSM and LM. The telescope was an integral part of that system. We spent the afternoon describing our Martin study and the possible added value of redesigning the telescope not only to take star sightings but to scan the surface and accept a hand-held camera the astronauts might carry so that the full surface area of the landing site could be photographed from within the LM.

The MIT engineers thought this would be possible but required a major design change. They were already experiencing some difficulties in meeting contract objectives, and asking for a change that might never be needed went beyond our pay grade. I wrote a report of our visit and drafted a memo to George Mueller, for Dr. Newell's signature, requesting that periscope modifications be considered. I don't know how this request was processed in OMSF; however, the modifications were considered too extensive and costly and the matter was dropped. Fortunately such an instrument was never needed on any of the Apollo landing missions.

With the Martin contract underway I started planning for several other studies. The Sonnett Report made it clear that a geophysical station capable of supporting five or six different experiments would be required. A drill that could extract core samples from as deep as possible below the lunar surface was another piece of equipment that we believed the scientific community would eventually call for.

Compared to Apollo, where we were told there would be constraints on all the important exploration parameters, we could think big. Most important, we could plan on taking very large payloads to the Moon's surface. Instead of 250 pounds we could think in terms of the 7,000 pounds or more mentioned earlier which could be used for any need we had. Experiments, life support and transportation headed the list of items that we defined to take advantage of larger payloads.

After studying USGS geologic maps of the Kepler and Copernicus regions made by studying the Moon by telescope, traverses of tens of miles seemed necessary to fully understand large lunar craters up to 50 miles in diameter. The astronauts would need mobility of some type to extend their activities beyond their immediate landing site, and the more capable we could make a vehicle the more useful it would be. Our limited understanding of the ongoing designs for the astronaut's space suits and life support backpacks indicated they would probably not be allowed to make long traverses on foot. Longer excursions would require a vehicle with a pressurized cab and full life support.

As with Mercury and Gemini, safety of the astronauts was always uppermost in our thoughts. We also needed to efficiently use astronaut time. Lunar surface tasks would be designed to take advantage of the ability of the astronauts to accomplish those aspects of exploration that humans do best: observe, describe, operate complex equipment, and respond to the unexpected. We did not want them performing manual labor that could be avoided. But we had to strike a delicate balance between automated and manual activities; otherwise supporters of unmanned exploration, both inside and outside NASA, would raise many questions and objections. Why go to the expense, not to mention risk, of sending astronauts if all they did was turn a switch and let a machine do the work? Switches could be turned on and off from Earth.

Our office never thought this a real challenge since the astronauts' unique abilities to observe and react to unexpected discoveries would always be their most important contribution toward exploring the Moon. A combination of automated equipment and hands-on tasks would be needed, and we took for granted that exploration would proceed in this fashion.

After starting the Martin study my next priority was to find a contractor to provide an overall analysis of how to design experiments and equipment for the post-Apollo missions. This study would generate estimates of weights, volumes, data transmission and power requirements for a suite of instruments selected by Foster's office. It was my first attempt at writing a government Request for Quotation (RFQ), and I received help from the NASA Headquarters Procurement Office. The RFQ was called *Scientific Mission Support Study for ALSS* and focused on the science operations that could be done from a mobile laboratory carrying two astronauts. It was released in early 1964 from our Headquarters office.

While writing this RFQ it became clear that managing contracts from Headquarters would be difficult in view of the large number of studies that we wanted to get under way. We hoped to find a NASA Center that would agree to manage them, which would have another positive attribute. It would be a strong voice supporting our ideas when budget time rolled around and we competed for scarce funds.

My few brief encounters with the MSC staff had not been very encouraging. At the time they were focused on Gemini missions and their initial Apollo science planning had not been accepted. What should be done after Apollo was definitely not on their agenda. This was brought home after Evans, Andrews and I briefed them on our post-Apollo planning and their reaction was a turned-up nose. In early 1964 I could not identify anyone at Houston who had the right background to manage the studies that I believed were needed.

Goddard Space Flight Center had built a strong Earth sciences staff that could have taken on these studies; however, they reported to OSS, the wrong part of NASA. The Kennedy Space Center, although a Manned Space Flight center, had as its primary responsibility servicing a variety of launch vehicles and did not, to my knowledge, have many Earth scientists on staff.

That left the Marshall Space Flight Center as my only choice and, as it turned out, a most fortuitous one. The studies initiated by our office and others in advanced missions to understand how to improve the Apollo hardware had been undertaken by several Marshall organizations.

Wernher von Braun, the German rocket genius brought to the US at the close of WW II, was the newly appointed Marshall director. He had just been reassigned from the Army's Redstone Arsenal at Huntsville, Alabama. At the end of the war the Army brought more than 120 German engineers and scientists to the US. Led by von Braun; they were directed to improve US rocket know-how. Of this original group some had been assigned to Cape Canaveral as well as Huntsville. With a perfect launch record at the Cape for their rockets and successful placement of the first US satellite in orbit, rocket technology was progressing at a rapid pace. Sending men to the Moon was their next challenge, which included building the huge new Saturn V.

Marshall was NASA's largest center in terms of staff, and I had had no previous contact with it. However, Ernst Stuhlinger, von Braun's science advisor on his original team, had been responsible for writing some of the reports recently made for our office. After several phone calls I scheduled a meeting in early 1964 with Jim Downey who managed Stuhlinger's Special Projects Office. Our first meeting was marked by some careful bureaucratic dancing reflecting his center's and immediate boss's cautious Germanic approach to having someone from Headquarters ask for a commitment of manpower and resources,

Jim, a University of Alabama graduate, was an easy-going fellow who commanded the respect of his unusual multitalented staff of scientists and engineers. My immediate impression was that they had the mix of skills needed to monitor the wide range of contractor studies I wanted to perform. Jim was eager to take on this new job, for his office had never received much funding for the earlier work Marshall accomplished for our office.

Jim wanted to know if my request represented a formal Headquarters assignment of new duties for Marshall. I didn't have the authority to say yes to such a pointed inquiry. I hedged but assured him that our office had funds available to support the studies that I was asking him to manage. I told him I would go back to Washington and start the needed paperwork. This meeting was the beginning of a long and productive relationship with Ernst Stuhlinger, Jim Downey and their staffs as we undertook a number of studies that broke new ground for lunar exploration. And from the very first day I knew that our personalities would mesh. We became good friends as well as working colleagues. Contracts were soon released from Marshall to study exploration of the Moon. The RFQ I started in Washington was also transferred to Marshall.

US Geological Survey Joins Our Team

As we started our studies at Marshall I began to build a strong partnership with the USGS office that Gene Shoemaker had just established at Flagstaff, Arizona. Gene was an outstanding scientist who eventually had a major impact on the Apollo program. After leaving Washington in the fall of 1963 he returned to Flagstaff, where he had recently moved with his wife Carolyn and three children. He had chosen Flagstaff for his new office for a number of reasons. It had a nice small-town atmosphere and was close to many interesting geological Moonlike features, all within an hour's drive to the east. Another plus, although Gene would have denied it, was that Flagstaff was far enough off the beaten track that he could run the show undistracted by his superiors in Washington. Over the years we had many private discussions about the disagreements he had with his bosses because he liked to cut corners on issues such as speeding up new hires to perform all the work he was being asked to do. The work he was beginning at Flagstaff would be very different in many ways from that conducted by the USGS during its long history as a government agency.

However, the local geology and structures were the real magnet. One of these, Meteor Crater, whose origin Gene had helped unravel, was about to become a "star" in the geological firmament, a place that all the astronauts would visit and study. He may have thought that in Flagstaff, three hours from Arizona's largest city, Phoenix, he could go about his business quietly. But knowledge of what was happening around the little town grew as his responsibility for training astronauts became known. In 1964 anything the astronauts did was sure to attract a large contingent of media and "hangers-on."

Although Gene's time in Washington had overlapped my arrival, we had never met. It soon became clear that I had to meet him. I learned more about the many activities in which his staff was already involved in addition to training the astronauts to be lunar explorers. It appeared that there could be a good match between his interests and my office's future studies. We talked several times on the phone about the direction in which post-Apollo planning was heading and agreed to meet to determine if he could support my needs.

In March 1964 I made my first visit. Gene's offices, in several one-story cinder block buildings at the Museum of Northern Arizona, were not very imposing. Furniture was very basic and looked like Army surplus. Some of the more innovative staffers had constructed packing box bookcases. I would spend many hours during future visits in Gordon Swann's small shared office; the only extra chair was a short plank he had set on his waste basket. But in spite of appearances I could sense the energy and dedication of the staff Gene was assembling; they hadn't come to Flagstaff for fancy accommodations.

Gene introduced me to those present, mostly young, some recent college graduates, and gave me a short tour. He had already been selected as a co-investigator for Ranger and the upcoming Surveyor program, our first attempts at obtaining close-up photos and measurements from the lunar surface. Some staffers were busily analyzing the first Ranger pictures returned only four months earlier and preparing for the first Surveyor landing. In addition to the Ranger and Surveyor work, Gene's office had the lead in making geologic maps of the Moon using large telescopes in Virginia and California.

Our meeting went well, and we agreed that we should work together on post-Apollo mission planning. The surrounding topography and geology would provide ideal sites for testing ideas on how to conduct long stay-time lunar missions, and it was obvious that Gene and his staff were passionate about being involved in exploring the Moon. He assured me he could hire additional staff for this new work. We made a handshake agreement to develop an interagency funding transfer, and I went back to Washington to start the paperwork. Eventually our handshake led to almost $1 million a year in cooperative work covering all aspects of post-Apollo lunar exploration.

Gene's staff involved in telescopic mapping the Moon's surface used their knowledge to plan geological exploration on the Moon. Most of the mapping work was being done at USGS offices in Menlo Park, California, using the nearby Lick Observatory telescope. Some Flagstaffers commuted to Menlo Park to work on their assigned maps. In mid-1964 the commute was shortened to a few miles when NASA built a 32" reflector telescope on Anderson Mesa, just south of Flagstaff, dedicated to providing geologic maps of the Moon. To complete their maps staffers spent many cold nights sitting at 8,000 feet peering through a telescope's eyepiece.

By the end of 1966 Gene had a major piece of many of NASA's pies: Ranger, Surveyor, Lunar Orbiter, lunar geologic mapping, astronaut training, principal investigator for the first Apollo landing missions, and post-Apollo science planning. At the height of his efforts, in 1968, over 190 USGS staff and university part-timers were working at several locations in Flagstaff, including a new government complex north of town. He would receive almost $2.5 million a year from NASA to fund his many assignments. Even his Washington bosses grudgingly agreed that Gene was improving the image of the USGS with the many favorable news stories of his work.

Using Photography from Earth Orbit

As mentioned earlier, Paul Lowman worked with the astronauts to have them photograph the Earth during the Mercury and Gemini missions. Paul believed that important geological insight could be gained from the photos taken during these missions and had compiled a long list of geological features to be photographed. Several on his list were suspected to be old impacts which would be important in understanding impact features on the Moon.

Each Gemini photo typically covered an area of some 3,500 square miles. Oblique photos covered even more, an unprecedented continuous view of the Earth's surface. Compare these numbers with the typical aerial survey whose average frame might cover less than ten square miles. Conventional photographic coverage of the large areas included in a typical Gemini frame would require construction of photo mosaics, the piecing together of many separate photographs by trained photogrammetrists. We knew that no matter how skillfully fabricated, finished mosaic maps could mislead a geologist. Features that looked like a stream or valley, or some geological feature such as a fault, might be artifacts of the mosaic process.

Features never fully photographed before the Gemini missions, such as the Richat structure in Mauritania and the Vredefort Crater in South Africa, were suspected impacts and of special interest not only as possible training sites but as opportunities to learn more about impact processes. In 1964 only a small number of well documented impact craters were known throughout the world. Some were so obscured by erosional processes that they were not well suited as training sites. Thus, we were constantly trying to find additional examples that we could study or use to train the astronauts. The only really good and accessible example was Meteor Crater east of Flagstaff.

Searching for Terrestrial Impact Craters

In 1964 I made two trips, one to Colombia, the other to Peru, to study possible impact structures seen on early Gemini photos. Paul accompanied me to Colombia to visit Lake Guatavita, a perfectly round lake located high in the mountains east of Bogotá. At the time of the Spanish conquest it was rumored to be a ceremonial site to which the Muisca Indian chief made a yearly visit to offer sacrifices by throwing gold pieces into the lake. Based on that rumor the Spanish attempted, unsuccessfully, to drain the lake but did dredge up several gold artifacts. In later years a few more were found and are in the Gold Museum in Bogotá. I knew of the lake, having visited it with scouts in my Boy Scout troop who were ready to take their first-class swimming tests.

Paul and I spent a week at the lake taking measurements and collecting samples for thin section study. We also studied the local geology to see if there were any geological reasons that would explain how the lake had formed in such a perfect circular shape. Returning home we studied our samples and couldn't find any indications that the lake's surrounding rock outcrops had been deformed by an impact. However, our study of the available literature led us to conclude that the lake had probably been formed when the ground collapsed over a dissolving salt diapir (a column of salt squeezed up from a thick subsurface salt formation). Some were known to exist in the area.

Undeterred by this failure, in 1968 I scheduled a trip to Peru where photos taken during Gemini missions showed a very large circular structure high in the Andes south of Cuzco. For this trip Paul wasn't available but a NASA colleague, Rollin Gillespie, offered to join me, paying his way and that of his son, a geology student at Stanford. Before leaving for Peru I had made arrangements to meet in Lima with the Peruvian Geological Survey and geologists with the Madrigal Mining Company in anticipation that they would provide support during the expedition.

I arrived in Lima without my baggage (it arrived two days later), and Rollin and I spent the next days meeting with Peruvian authorities and mining company officials discussing the geology near the structure and arranging transportation and logistics. Rollin's son and a friend would be coming a few days later, and we agreed to meet them and the accompanying Peruvian geologists at the town nearest to the structure, Sicuani. They would bring four-wheel-drive vehicles and other equipment. Rollin and I then flew to Cuzco, where we met with staff from the National University of San Antonio and explained our objective. Their geologists were unaware of the structure and surprised to see it so clearly shown on the Gemini photos. I promised to provide a lecture for them on the Apollo program on my return from Sicuani.

Off Rollin and I went on a local bus that serviced Sicuani, about 75 miles south of Cuzco and halfway to Lake Titicaca, We were accompanied by many passengers and their luggage, some of the four-footed and two-footed winged variety, carried inside and outside the bus. Soon after leaving Cuzco the road turned into a rough, narrow, winding gravel track following a route that skirted deep canyons. Off in the distance, from time to time, we caught glimpses of snow-covered mountain peaks. It was an uncomfortable ride with frequent stops but an interesting introduction to the high Andes. We arrived in Sicuani late in the afternoon and found lodging at the hotel recommended by the university staff, a basic accommodation with very thin mattresses, no heat in the room and warm water promised for a short time early in the morning.

At over 12,000 feet Sicuani was a new South American experience for me. It was very cold walking around the main plaza that evening but everyone was friendly and we drew curious stares, Rollin at six feet four and I at almost six feet two. The locals all seemed of indian descent, very short, wearing serapes and broad brimmed hats of various shapes. My most memorable experience that night was looking up into the sky and seeing from horizon to horizon the Milky Way, undimmed by the few competing lights in the town. It appeared to be a dense, broad carpet of brilliant stars, large and small and so clear it seemed as if I could reach up and touch them. There were no twinkling stars as there was little intervening atmosphere to make them twinkle. Now I could appreciate why astronomers from many countries wanted to locate their instruments in the Andes.

That night in the plaza we met an American Carmelite priest who, when he learned where we were staying, invited us to spend the next days at the bishop's parish house with hot water all day. After shivering in bed the first night still in my street clothes (whoever thought about packing warm pajamas when traveling in the tropics?) and under a thin blanket, Rollin and I gratefully accepted the invitation and moved all our gear to the bishop's house the next morning. When we met Bishop Hayes he turned out to be a man of many talents, including a ham radio operator. He frequently communicated with a ham in northern Virginia and called giving him our home phone numbers so he could call Pat and Rollin's wife to say we had arrived safely.

Hayes was very familiar with the area and, when we discussed our plan and objective, said it would be impossible; even four-wheel-drive vehicles could not get up in the mountains. The only way was on horseback. That was an unexpected setback; the Peruvian geologists in Lima had told us we could drive close to our study area. Adding to that problem, Rollin's son and the rest of the party had been held up in Lima getting the vehicles and would arrive a few days late. From my previous Colombia experience and the problems one could expect to encounter when working in Latin America, I thought I had all the bases covered, but these problems went beyond my planning. Before the expedition was ready to start, including renting horses, the time I had allocated for my stay would be up and I would have to return to the States, never getting to the "crater."

With some misgivings I sketched out a plan for Rollin and his son to follow and returned to Lima via Cuzco, giving the promised lecture at the university, then back home. When Rollin returned to NASA he brought only a few samples and none from near the "crater." It had been almost impossible to travel in the mountains, even on horseback. On examination we could not find any indication that an impact had formed the structure. Perhaps the sampling had not been sufficient in view of all the problems, but as far as I know the question of how the structure seen on the Gemini photos was formed is still unanswered.

A few weeks after returning from Peru I received an envelope from Professor Carlos Kalafotivich at the university in Cuzco. It contained several Peruvian newspaper clippings noting that NASA scientists had visited the region and were interested in the mountains near Sicuani. The papers featured bold black headlines that translated to read "Flying Saucers Land in Canchis" (a small town near Sicuani). Local people interviewed said they were very familiar with the mountains and had observed frequent visits by flying saucers. The saucers came, they said, so that aliens could extract precious gems from the mountains and take them back to their home planet. Now I knew what had influenced me to visit Sicuani.

Despite our lack of success in Colombia and Peru, Paul and I continued to study the photos after the last Gemini mission in 1966 in the hope of finding terrestrial analogues that could be used for post-Apollo astronaut training.

Developing Apollo Science Experiments

Planning for Apollo science finally began in earnest in 1965, and I became closely involved. Three years had passed since the last National Academy of Sciences summer study that led to the Sonnett Report. Since then we had learned a lot. The close-up views of the lunar surface taken by the Ranger spacecraft increased our confidence that "normal" geological studies could be planned for the astronauts.

July 1965 was selected as the date for the National Academy Space Science Board to meet at Woods Hole, Massachusetts, and review the status of space science. Dick Allenby and I planned to take advantage of the assembled experts to review our progress and receive specific recommendations for Apollo and post-Apollo science experiments and operations. We set up a separate meeting for invited space scientists and engineers at the Falmouth high school for the same date so that the Academy attendees who desired could join our sessions and we, in turn, could benefit from any of the Board's new recommendations coming out of their meeting.

Working groups were established in eight scientific disciplines: Geology, Geophysics, Geodesy/Cartography, Bioscience, Geochemistry, Particles and Fields, Lunar Atmosphere Measurements, and Astronomy. This last discipline was added at the eleventh hour to review the preliminary findings of our study being conducted at Marshall to include a telescope on post-Apollo missions.

The Astronomy working group was asked to look beyond Apollo to lunar bases where the Moon could become the site of large observatories that might include radio telescopes on its far side to shield the instruments from Earth-made electronic noise. In 1965 no astronomy experiment was planned for any of the Apollo missions. However, a far UV camera-spectrograph telescope was eventually included on the Apollo-16 mission.

Each working group submitted a report summarizing the results of its deliberations. The combined report, *NASA 1965 Summer Conference on Lunar Exploration and Science*, immediately supplanted the Sonnett Report as the authoritative reference for Apollo and post-Apollo lunar science planning.

For the next four years I spent most of my time working with all the different stake holders involved in planning and developing Apollo experiments. Post-Apollo planning continued, and some of the experiments and equipment we proposed for those missions eventually found their way into the Apollo program. Science became more and more an important part of Apollo planning, a big step forward compared to the situation when I arrived in 1963. Geological tools, various cameras, and a nuclear-powered geophysical station were developed. Science instruction and field trips became an integral part of all the training the astronauts endured.

We convinced Chris Kraft that Mission Control should include a science support room as part of the team monitoring the astronauts' actions on the lunar surface. This major concession by Houston's hard-nosed engineers, who were focused on assuring astronaut safety, came only after a long and sometimes contentious debate. Including complicated science tasks for the astronauts to perform was a late addition to the already full schedule of activities they would perform after they landed.

By 1967 100 hours of classroom lectures and ten field trips became the requirement for astronaut geology training. This training and mission simulations would become more and more rigorous and realistic as the program matured. Simulations were scheduled utilizing prototype and final design equipment and tools. All the astronauts participated at some level in both classroom and field training. The first three astronaut selection groups had the most extensive training, but no one knew who would ultimately be selected, so we tried to have all the astronauts at as high a level of competence as possible in their geological training within the constraint of time available. Many famous geologists and geophysicists volunteered their time to assist in the training. Some stayed on to become members of the Apollo Field Geology teams, working with the astronauts until the last mission, Apollo-17 in December 1972, was safely home.

Commercial Application of Gemini Photos

A few of the Gemini photos mentioned earlier had been published in the *National Geographic, Life Magazine*, newspapers and other publications, but the vast majority had not been seen by the general public. In his spare time Paul had carefully cataloged the pictures and interpreted their geologic content. It occurred to us that these new views of the Earth might be of interest to companies conducting exploration in remote parts of the world. Thus far no commercial interest had been shown in the Gemini photos. A positive commercial response would support NASA's proposed Earth orbital remote-sensing program, in an early planning stage by Peter Badgley, and perhaps convince NASA management to accelerate the program.

In May 1966 I called Mobil Oil in New York and talked to an old boss in Colombia, Jim Roberts, who had been transferred first to Venezuela and then to Mobil headquarters. I explained what we had and what we thought would be the potential benefits and application of space photography. He was interested in seeing the photographs and agreed to set up a meeting with some of the Mobil exploration staff, the unit responsible for finding new oil fields.

A few weeks later Paul and I flew to New York to show the Gemini photos to their first commercial audience. We brought to the briefing some of the best examples of geological features photographed by the astronauts, mountain ranges in the southern Sahara and structures in Iran of the type petroleum geologists looked for (anticlines and synclines). I knew that Mobil was exploring in Libya and Tunisia, and had field parties working in those countries and had begun exploration in Iran, possible destinations for me if I had stayed with Mobil. We also included a few spectacular views of the Andes and Himalayan mountain chains. We felt sure that no aerial photographs existed of some of these areas, and this would be the first time Mobil staff had seen such views. We thought they would be impressed.

We were wrong. For unknown reasons the staff that Roberts brought to our meeting showed little interest. They indicated that they had, or could obtain, sufficient conventional coverage so that space photographs were not needed. This response mystified us. Perhaps they felt that an endorsement would leave them open to providing financial support for an undertaking with an uncertain future. We will never know what might have happened if Mobil had been enthusiastic. Like other programs struggling to get started, the Earth orbital observation program limped along, in part because of the lack of strong commercial interest. It would be many years before Landsat photos would be available and of great interest to many different commercial users.

Astronaut Application

Although I had been a military pilot, as were all the astronauts in the first selection groups, I didn't have a lot of jet hours, mostly prop time logged many years earlier. After working with the astronauts and knowing of their flight backgrounds, I could see that it would be virtually impossible for me to qualify in a typical selection process because of my lack of current pilot experience. Many of the astronauts had been test pilots, a position I had once attempted to qualify for but was denied. Test pilots had obvious qualifications that made them excellent astronaut candidates.

Gene Shoemaker had lobbied for a scientist-astronaut selection and, before he was diagnosed with Addison's disease, had been considered a probable top candidate when NASA finally agreed it needed such positions. Even when he knew he could not be selected he continued to lobby along with many others in the science community. Knowing that there was a strong possibility that NASA would recruit astronauts with strong science backgrounds, especially in the Earth sciences, I bided my time, feeling that my best chance to qualify would be through such a program. My patience was rewarded; in 1964 NASA asked the National Academy of Sciences to develop procedures for a scientist-astronaut selection.

When the opportunity to submit an application was finally announced in October 1964, I quickly obtained the packet that included all the paperwork needed in order to apply. Many qualification standards were listed, including age, height and educational background. Height! Maximum allowed height was 6 feet. I was 6'1" plus. The age limitation excluded anyone born before August 1, 1930. I was nine months too old. I made a few calls to determine if these were inflexible requirements; they were. The height restriction was based on the dimensions of the Gemini capsules and the Apollo equipment then under design. It was projected that anyone over six feet would not be comfortable in the spacecraft.

With great disappointment I wrote a letter to the Academy, the initial screening hurdle, to tell them I was interested but disqualified because my age and height were slightly over the announced standards. I expressed hope that these restrictions might one day be changed so that I and others in my predicament could apply. I wrote my letter with deliberate forethought. I felt sure there would be other scientist-astronaut selections. Our post-Apollo planning called for extensive scientific activities on the lunar surface that only trained scientists could perform. George Mueller had testified before Congress on these plans, and I knew he supported the need for additional scientist-astronauts. I hoped my letter would be on file and retrieved at the next selection, perhaps influencing the selection criteria and definitely showing my long-term interest in the program.

To give myself a better chance in the next selection, whenever it would be, in 1964 I applied for a pilot slot in the Navy Ready Reserve squadrons that drilled at nearby Andrews Air Force Base. No pilot openings were available. Instead I joined an intelligence unit drilling once a month to get back in the Ready Reserve flow and learn through the grapevine where pilot assignments might be found.

I soon learned of a reserve pilot vacancy at NAS Lakehurst in New Jersey. I quickly transferred to VS-751, an anti-submarine squadron, and resumed flying again after a seven-year layoff. A year and a half later, with new flying time under my belt, I convinced a fighter squadron commander at Andrews to ask for my transfer to his squadron. I began the transition to the F8U Crusader. But the Navy got wind of this behind-the-scenes activity. Apparently needing anti-submarine pilots more than fighter pilots, that transfer was rescinded and I was assigned to VS-661 at Andrews. I was disappointed since I was really looking forward to flying the Crusader, one of the Navy's best ever fighters, but the transfer had one redeeming factor. I would now fly the S2F out of Andrews and save the long monthly commute to Lakehurst. Although I wasn't flying jets at least I was flying and hoped that this would be a plus in the next selection.

In September 1966 the second scientist-astronaut selection was announced by the National Academy of Sciences. Accompanying the press release was a short statement by Shoemaker, who would be the chairman of the Academy's selection panel: "Scientific investigations from manned space platforms and direct observations on the Moon will initiate a new phase in man's quest for knowledge. While such missions call for daring and courage of a rare kind, for the scientist they will also represent a unique adventure of the mind, requiring maturity and judgment of a high order." Who could resist such a challenge? With Gene as chairman and knowing several other members on the panel that would review the applications, I thought I would have a real chance to be selected.

It was rumored that this class would be larger than the six selected in 1965, thus increasing my odds for selection. The number of applications received by the Academy in 1964 had been somewhat disappointing. As a result the criteria were a little more relaxed this time. Age and height limitations had not changed, but the press release stated that "exceptions to any of the requirements will be allowed in outstanding cases." Perhaps I could qualify as an outstanding case based on my pilot experience and knowledge of lunar exploration.

My application must have been one of the first of the almost 5,000 received at the Academy. Evidently there had been enough good publicity about the Apollo program to encourage many young scientists to be a part of it. About 200 were selected by the Academy for the next phase of physical and psychological examinations, and I made the cut. We were divided into small groups and sent to the Air Force School of Aerospace Medicine at Brooks Air Force Base in San Antonio, where all astronaut candidates were screened.

We endured one week of prodding, blood taking, spinning, IQ and many other tests, some of which were vividly shown in the movie *The Right Stuff* but without the comic detail. While being tilted upside down with my stomach filled with a barium solution, it was discovered that I had a slight hiatal hernia; my esophagus didn't close tightly enough to hold all of the solution in my stomach. Because it was apparently a very minor ailment and because, I assume, the other test results were quite good, I was sent to Walter Reed Medical Center in Washington, D.C., for a second opinion. The examination at Walter Reed went well. The examining doctor wrote a great letter back to NASA (he gave me a copy) saying that he did not consider the diagnosis disqualifying. At worst I might have to take an antacid to relieve any discomfort I might feel while on a mission in zero G.

Where did this leave me? I could not be sure, but I had enough experience to know that astronaut selections were made in a secretive fashion and that Deke Slayton and Al Shepard were undoubtedly involved. Who else I didn't know.

By now I was acquainted with all the astronauts, including Al and Deke, but I wasn't sure if this was good or bad. I had been on field trips with them and from time to time was invited to brief them on how the plans for post-Apollo missions were proceeding. I was often in the astronauts' office building to visit them. After Jack Schmitt was included in the first scientist-astronaut selection we played racquet ball together in the astronauts' private gym at MSC. I felt I had a good relationship with all that I had met. But perhaps my differences with some MSC managers involved in developing Apollo science might hamper my selection. MSC was a very tightly knit organization. In June I received the call I had been hoping for. I had made the final cut and was invited to Houston for the last interviews before a selection was made.

Twenty-one candidates made this final visit. I knew a few of them from my week in San Antonio. Their backgrounds included almost all scientific disciplines, but as I read the list I saw I was the lone geologist along with one geophysicist. Only two earth scientists! Most of the post-Apollo science activities we were planning had some earth science connection. I thought my selection was in the bag.

After arriving at MSC the first scheduled activity was a ride in a T-38, the astronauts' aircraft of choice, based at Ellington Air Force Base. This was a piece of cake. I flew the plane from the front seat with a NASA pilot (perhaps evaluator?) in back. We did some simple maneuvers, a few snap rolls, and I generally showed off my flying skills. From what I read in the bios of the other candidates I was the only one with jet pilot experience. If this was a test, I must have passed.

Next I took a ride in the MSC centrifuge that spun me up to about 6 Gs while performing a few simple exercises of hitting some light switches. This was not a problem, and I suspect potential future bosses were looking on through closed-circuit television to see how I performed during the nearest thing to a stress test.

After a few other briefings came THE interview. Only four people were in the room, Al, Deke, Bill Hess representing MSC's science side, and Chuck Berry. Berry headed the medical sciences office and was more commonly known as "the astronauts' doctor." All the questions were rather innocuous. Berry asked about the hiatal hernia and, since I had seen the Walter Reed report, I told him I hadn't even known I had it until the test and thought it would not be a problem.

The only question that stands out in my mind was the one asked by Deke: "Don't you think you're too old to be an astronaut?" I was 37 but knew I wasn't the oldest of the final 21 candidates. However, as I was over the advertised age allowance, I had done a little homework. I answered, "I don't think so. I'm younger than Wally Schirra, and he's still flying." This brought a big laugh from all four inquisitors. Considering that Schirra, then 43, was the only astronaut from the original seven to fly in all three programs, Mercury, Gemini and Apollo, my answer was evidently on the mark. That ended the interview, and Al said he would give me a call. I thought my selection was only a formality. That afternoon I did some preliminary house hunting in the neighborhoods around MSC.

In August Al called. "Don," he said, "I'm sorry to tell you you weren't selected." We talked for a few more minutes and I'm sure he sensed my disappointment. They had selected eleven in the scientist-astronaut class of 1967, including the only other Earth scientist, Tony England, the geophysicist. I didn't ask Al why I wasn't selected, as I was sure he wouldn't give me any specifics. I rationalized that it was a combination of things. They didn't want to take any chances with my hiatal hernia. Why need to add some Tums to tight payloads? My GS-15 rating at the time would have made me senior to most of the other astronauts from a government classification standpoint. My pilot background may have been a negative; I would have been the only one who would not need pilot training. What would they do with me during the year the others were in flight school? And, perhaps, there may have been some negative comments from MSC managers with whom I had disagreed in years past.

Al died in 1998, and I never had a chance to ask him why I wasn't chosen. Perhaps if I had the chance to ask him many years after the 1967 selection he would have told me and perhaps not. Probably he wouldn't have remembered. In any case, my non-selection probably did me a favor from a career standpoint. Within two years the post-Apollo missions, the prime reason for the 1967 selection, were canceled, and none of the 1967 class flew on a space mission for 15 years. Joe Allen was the first from this class to fly as a mission specialist on shuttle flight STS-5 in November 1982. A few retired or left NASA before participating in any NASA missions.

But the good news of the scientist-astronaut selections was that Jack Schmitt, selected in the 1965 class and who had been working on projects I was sponsoring at Flagstaff, was at MSC. He became a strong advocate among the astronauts to include science experiments on the missions and saw eye to eye with our concerns that some MSC managers were attempting to control Apollo science. However, members of the scientist-astronaut class of 1967 played important roles in training during Apollo missions and were CapComs. Joe Allen, Bob Parker, Story Musgrave and others stayed on, making major contributions to

NASA programs after Apollo was completed. And, as the final icing on the cake, Jack was on the last landing mission, Apollo-17.

Soviet Lunar Space Missions

Although the USGS was studying data in the '60s coming back from the Ranger, Surveyor and Lunar Orbiter missions, it had been a game of "catch up" with the Soviets. Their Luna-2 was the first unmanned probe to impact the Moon in September 1959. Two months later Luna-3 flew behind the Moon's perpetually hidden far side and for the first time took a few pictures, a real coup. After that date we traded sending unmanned spacecraft to the Moon; some, on both sides, did not achieve their objectives. In December 1966 the Soviets made the first successful soft landing of a spacecraft on the Moon that sent back close-up pictures of the lunar surface. Our Surveyor-5 would not duplicate that feat until September 1967. However, the data returned by Surveyor-5 had greater scientific value as it included important experiments that measured some characteristics of the lunar surface.

We were envious of the Soviets' successes, but an up-side to their accomplishments was revealed to me. Deep in the bowels of NASA Headquarters, protected from prying eyes and ears, was an office that provided liaison with other government agencies involved with collecting classified information on Soviet space programs. That office was managed by fellow NASA staffer and friend, Myron Krueger. As I held a secret clearance Myron kept me up to date, to the degree he could, on what the Soviets were doing in addition to the information released to the media. I would have needed an even higher clearance than secret for Myron to tell me all he knew.

Nevertheless, when the Soviets successfully landed Lunokhod-1 on the Sea of Rains in November 1970, he told me for the first time about one of our surveillance capabilities. We were intercepting all the transmissions the Soviets were receiving from the Moon. Soon after the landing he called me over to his office and handed me a thick red album. Inside were nine panoramas made from 36 individual pictures taken as the Lunokhod moved a short distance from its original landing point. Moreover, he assured me that the detail on the photos in the album was significantly better than what the Soviets were looking at. Our receiving technology was superior to theirs. The photos of the lunar surface were very clear, and we used them as we prepared for our final missions. This same office was involved in the sudden decision to send Apollo-8 to the Moon at Christmas 1968 after our surveillance showed the Soviets preparing to launch their largest rocket. Now you know the rest of that story.

Preparing for the First Landing

In December 1967 the last major organizational change involving Apollo science was made at NASA Headquarters. Mueller established the Apollo Lunar Exploration Office reporting to Sam Phillips. A month earlier NASA had received the news that there would not be any further appropriations provided for post-Apollo lunar mission planning so that activity was now in limbo. Mueller put Lee Scherer in charge of the new office. Lee had just finished wrapping up the loose ends left over from the very successful Lunar Orbiter program, and this appointment gave him a chance to expand his management role. His new office combined the responsibilities of Foster's office that reported to both OSS and OMSF and some of the post-Apollo lunar exploration activities of Manned Lunar Missions. Still wearing my two unofficial organization hats but officially assigned to the Manned Lunar Missions office, I was transferred to Scherer's new organization. Scherer also inherited most of Foster's staff and other NASA Headquarters staff who were involved in lunar science, including those managing the development of ALSEP, the Apollo geophysical station that had been under the direction of the Office of Space Science.

Scherer's appointment was a master stroke by Mueller. Scherer was well liked and trusted by the science side of NASA, having managed the Lunar Orbiter program, an OSS responsibility. As a result of the close connection of Lunar Orbiter to Apollo for landing site selection, he was well known to OMSF management. After our initial meeting in 1963 I got to know him well, working with him and his NASA and contractor teams on Lunar Orbiter site selection. Perhaps it was the Navy connection and my familiarity with his Navy way of doing business, but now I expected to see more progress being made in all aspects of Apollo science. Lee would have a much greater impact on Apollo decision makers than had Will Foster.

Being on Phillips' staff put Scherer directly in the Apollo chain of command. Many of the senior NASA managers on Apollo were either active duty or retired military officers, so Lee fit right in. Those of us who had been working on Apollo science were no longer half OSS and half OMSF, with neither office quite sure whose side we were on. Now we were all clearly part of the Apollo team.

With my new colleagues I moved into the Apollo offices at the just completed L'Enfant Plaza complex, where we remained until the last mission came home. I was given the new title of Program Manager, Plans and Objectives, with responsibilities that permitted me to be involved in all aspects of Apollo science and, most importantly, the planning for what would be done after the first successful missions. We never doubted Apollo would be successful, thus we planned to take full advantage of the missions that would follow the first landing and conduct as much exploration as possible. But two-week missions and building a lunar base did not appear to be in the cards because those funds were not appropriated.

Antiwar Demonstrations

While we were focused on Apollo, many others were concentrating on a very divisive issue, the war in Viet Nam. Protests organized by overt anarchist and communist leaning organizations were erupting on campuses around the country. In the spring of 1968 Columbia University was the scene of a disgraceful demonstration that, since it was in New York City, was extensively covered by the national media.

It began rather peacefully but, egged on by Columbia Students for a Democratic Society (SDS) and some notorious fellow-travelers who came to the campus, it quickly turned ugly. The Dean's office was invaded and he was held captive. Classrooms were trashed. Some demonstrators marched to the NROTC Armory and trashed and burned it. New York's finest responded and the demonstrations were finally quelled after a few days. Pictures of the police swinging billy clubs were in all the papers and on TV, with some in the media claiming an unjustified overreaction.

The University capitulated to the demands of the rioters. One of the demands was to abolish NROTC at Columbia. It was agreed to by Columbia's administration, a sad footnote to the University's illustrious history that only recently had boasted of having a former five-star general at the helm who then became the nation's thirty-fourth president. The agitators then went to other campuses and later to the Democratic national convention in Chicago. Subsequent Columbia administrations never apologized for acquiescing to this outrageous demand and never permitted the NROTC to return to the campus.

Forty years later, when *Columbia College Today* (CCT) printed a series of letters on the anniversary of the riots, many supportive, some not, I contributed my take on the event. I condemned the riots and instigators. I was challenged in the next month's edition by another alum who clearly sympathized with the hooligans and SDS, stating that the associations and criticism I made were wrong. CCT allowed me to respond with facts that supported my allegations. They were not challenged in subsequent issues.

Back to Apollo

After the lunar orbit flight of Apollo-8 at Christmas, 1968, anticipation grew for the first lunar landing attempt. However, prior to the Apollo-8 flight NASA management began to reevaluate the first landing mission's lunar surface activities. There was a growing concern about how well the astronauts would function on the lunar surface and, more importantly, how the LM would perform during the planned two-day stay. The LM, experiencing several problems, was behind schedule and had not yet been tested in Earth orbit or during a lunar mission.

Several ways to alleviate these concerns were explored. First, the two planned EVAs could be reduced to one or the length of each shortened. However, if only one EVA was allowed, then the ALSEP deployment could not be accomplished and still leave time for the astronauts to carry out other important tasks, including sample collection and photography. Not carrying the ALSEP reduced the astronaut workload and the weight carried by the LM for the first mission, a partial solution to both concerns. By removing the ALSEP's weight a few seconds of additional hover time would become available. ALSEP became a prime target for removal.

When rumors spread that science activities on the first landing might be drastically reduced, Charles Townes, chairman of the Science and Technology Advisory Committee, urged NASA senior management to keep as much science as possible on the first mission. Our office and others in the scientific community were also lobbying hard to keep ALSEP on the first landing mission and to maintain two EVAs. We were fighting for more than just science. ALSEP and the geological activities the astronauts were scheduled to carry out represented years of planning and hard work, not to mention suffering through many contentious meetings with those in NASA who had not embraced the need for science on Apollo. We were not willing to accept a major reduction of science activities on the first mission. Who could predict the future? We might never get another chance to try to answer important questions.

In September and October, in response to this outcry, an alternative to dropping the ALSEP was studied by our office and MSC and presented to NASA's Senior Management Council. A new, much smaller and more easily deployed science package was approved and given the name Early Apollo Scientific Experiments Package (EASEP). EASEP would comprise just three experiments, the Passive Seismometer, the Dust Detector and the Laser Ranging Retro-Reflector. Another high priority experiment, Solar Wind Composition, was self-contained and easily deployed. The Field Geology experiment and sampling would constitute the remainder of the science payload.

EASEP would be much lighter than ALSEP and require less time to deploy. By including it NASA hoped to divert criticism and show that its science heart was in the right place. The astronauts would leave the highest priority geophysical experiment, the seismometer, at the landing site. They would still have time to conduct a limited geological study but would collect fewer samples than originally planned.

Instead of ALSEP's nuclear power source, EASEP would get its power from batteries charged by solar panels. Solar arrays had been rejected for ALSEP but now were acceptable for EASEP because of its probable short operating lifetime. EASEP's design would only guarantee that the seismometer would operate through the remainder of the lunar day in which it was deployed, about ten Earth days. EASEP included several small nuclear isotope heaters in hopes it would survive the cold lunar night when power wasn't being generated to transmit data. With a little luck it might resume functioning during the next lunar day when the solar panels would once again generate enough power. Like ALSEP it had a self-contained telemetry system to transmit to Earth the seismometer and dust detector readings.

After Apollo-8 only two test flights were scheduled before the first landing attempt. EASEP would have to be built in five months to meet a required May 1 Kennedy Space Center delivery date in order to be packaged on the LM for a June or July flight. And there had to be time available for the crew to practice deployment with a high-fidelity mockup. Built by Bendix, EASEP proceeded without a hitch and was delivered to KSC one day ahead of schedule.

The Flight of Apollo-11

On July 16, 1969, along with Pat's parents and a multitude of other sightseers (local Civil Defense officials would later estimate one million people), we were on hand to watch the launch of Apollo-11. We were parked along the shoulder of US 1 in our Winnebago camper about five miles north of KSC and the launch site. We had selected our viewing point the night before and thought we were fortunate to find a spot so close. It had been a madhouse trying to drive near the Cape; no one seemed to care about normal rules of the road as each car or camper vied for preferred spots and parked willy-nilly as inclinations dictated. Local and state police tried to maintain some order, but it was a hopeless job. In the early morning, as launch time approached we climbed on the roof of our motorhome to get an unobstructed view, meanwhile listening on the radio to Jack King, "the voice of Apollo," counting down the final seconds before the launch.

The Stars and Stripes were flying everywhere, and the crowd was in a party mood. The countdown proceeded smoothly, and at 8:32 AM the Saturn lifted off to loud cheers and many teary eyes, including mine. Our hearts went with the crew of Apollo-11. This was the second Saturn V launch I had witnessed, but I still wasn't prepared for the enormous sound and low-frequency reverberations that reached us, even at this distance, a few seconds after the Saturn had cleared the launch tower. We watched for several minutes as it disappeared to the east, leaving behind a huge plume of white smoke. Then we went inside, finished breakfast, and talked about what we had just seen and what lay ahead.

I was in a hurry to leave because I was due back in Washington to help broadcast the landing on Voice of America. It was almost an hour before the traffic jam began to move and we were on the road. Apollo-11 was on its way to the Moon with the first science payload that men would place on another body in our solar system. If all went as scheduled, Neil Armstrong and Buzz Aldrin would have the honor of being the first humans to set foot on another body in our solar system, making the first close-up studies of the Moon's surface and describing how it felt to be walking in one-sixth gravity. Mike Collins, the Command Module pilot, would be waiting in lunar orbit to rendezvous with the LM and, if necessary, lower his orbit in case the LM ascent stage did not perform as well as planned.

Four nights after the launch, in anticipation of a successful landing, the Voice of America (VOA) had assembled a team to report on this once-in-a-lifetime adventure for its worldwide audience. Several NASA colleagues and I were in the VOA Washington studios as "color commentators" to back up the VOA reporters, led by Rhett Turner, who would be reporting from the Manned Spacecraft Center in Houston. We listened anxiously with millions of others around the globe to the exchange between CapCom Charlie Duke and Armstrong and Aldrin in the Eagle as they went through the final maneuvers to land the LM.

The excitement of those last few minutes, heightened by the difficulties the crew endured in selecting their landing site as alarms rang in their ears and their fuel supply neared exhaustion, made Armstrong's announcement, "Houston, Tranquility Base here. The Eagle has landed," almost anticlimactic. We could hear the cheering in Mission Control through Rhett's microphone, and we in VOA's Washington studio were shouting and pounding each other on the back. Although we had worked for years to help achieve this moment, it seemed somewhat incredible that the first try was successful. Armstrong and Aldrin really did land and walk on the Moon, not on a movie set in the Arizona desert as some believe.

We were primed to discuss the mission in great detail, but as the night unfolded only a few questions were directed our way. I was seldom asked to show my vast understanding and insight into things lunar; VOA wasn't about to share much of the limelight on this historic occasion. I did, however, receive a card from some friends in Colombia who said they had heard me on VOA, expressed their great pride in Apollo-11's success, and congratulated me for being part of the program. I wondered if some of my former colleagues would remember their skepticism at my decision to leave Mobil six years earlier and join NASA. I certainly did not regret that decision.

Family Vacations - Father's Death

Before continuing with the unfolding Apollo story, I'll digress with a few family stories. It was always difficult to break away from my NASA responsibilities. To take best advantage of my free time we bought the Winnebago motor home mentioned above. It was 22 feet long and was advertised to sleep eight in somewhat close proximity. It came with a full kitchen, bathroom and shower, shag carpet and generator. We also carried a battery-operated TV that could be plugged into an outlet when the generator was running. It was ideal for short or long camping and sightseeing vacations. I used it during Cub Scout trips when I took several dads and sons camping along with the rest of the boys in the pack. The fathers slept inside and the boys in tents. It was an easy way to introduce our sons to camping at age eight.

I scheduled a two-week vacation in 1968 to visit Alaska. To get the most out of the short time Pat, her father and mother, the boys and our German shepherd Smoke drove in the Winnebago to Seattle. I flew out and met them there. We drove to Prince Rupert and took the inland ferry to Haines with two one-day stops at Wrangell and Juneau. From Haines we drove north to Mount McKinley, stopping each afternoon along the way at a lake or river to fish. In 1968 we were allowed to drive all the way to the base of the mountain, and we had great weather that allowed clear views with just a few clouds brushing the peak. In the valley below our camp site we watched caribou, bear and other wildlife roaming around unconcerned with our presence. From Mount McKinley we continued north to Fairbanks and visited with a friend who had been in graduate school with me at Mines. He was on the faculty at the University of Alaska. Aside from pesky black flies, we enjoyed wonderful weather while taking in Alaska's unmatched scenery completely free of a noisy crowded world. From Fairbanks we turned south, picked up the ALCAN Highway and drove back to Great Falls, Montana. I flew back to Washington and the family drove back to Maryland.

The ease of our trip in the motor home and the fun we had convinced me to trade in the first Winnebago for a larger 28-foot model. The additional six feet and a few more bells and whistles really made a difference. I also added a second 50-gallon gas tank. A year later on another two-week vacation in more luxurious accommodations we went to Mexico. Pat, her father and mother, my mother, the boys and the dog drove to Brownsville, Texas, where I met them. We drove south, finding safe camping spots each night, and stopped at Mexico City to visit the Gilmores, who now lived there after finishing an Ingersoll Rand assignment in England. We visited all the local archaeological sites and then drove to Merida on the Yucatan Peninsula. We visited the Mayan cities of Chichen Itza, Palenque and Uxmal and camped near those locations and on the beaches in solitary enjoyment, then drove back to Brownsville, where I left to fly back to Washington. The family continued at a more leisurely pace to Maryland. Motorhomes are an ideal way for families to see America.

Shortly before we returned to the States from Colombia my mother and father slipped their New Jersey bonds and moved to Bay Pines, Florida. They chose a home three blocks from the large Veterans Hospital. By this time my father, a two-pack-a-day Camels smoker, was experiencing some heart problems. My mother immediately joined the hospital volunteers and soon was in charge. She spent many hours there and was a favorite of the staff for the activities she organized and supervised for the patients. In addition, although a latecomer to driving (few women of her generation drove), she began to teach defensive driving.

We visited them several times in our motorhome. Their house was on a small lake, and my father, never one to sit idle, built a small cantilevered concrete pier that extended over the water. Tommy and Bruce fished from the pier. But Dad's smoking caught up with him, and after several serious hospitalizations he died in October 1969. My mother said he wanted to be with other veterans and I arranged for his burial on October 24, 1969, at the Barrancas National Cemetery at the Pensacola Naval Air Station.

Apollo Missions Continue

After Apollo-11's successful flight Harry Hess, chairman of the National Academy of Sciences Space Science Board, convened a meeting of the board on August 25, 1969, at Woods Hole, Massachusetts. During the Apollo program the board had provided advice and recommendations to NASA on the overall structure of lunar science and specific experiments that should be conducted. Harry was concerned that Congress and the public were already tired of supporting Apollo missions. NASA's budgets had steadily declined after Fiscal Year 1965, four years before the Apollo-11 flight. Harry wanted to take advantage of the Apollo-11 mission success and publicity to be sure the scientific community would not join the Congressional chorus to terminate Apollo but would actively support further lunar exploration.

Harry invited me to attend to describe the experiments and exploration program that we planned for the remaining Apollo missions. Each of the six scheduled flights would be scientifically more ambitious than the last. For the last three, you may remember, there was a small two-man vehicle in the payload that would greatly improve the astronauts' ability to study their landing sites. Over 50 experiments had been built or were being built specifically designed to take advantage of the Moon's unique environment and conditions where they would be deployed.

At the coffee break the first morning Harry said he wasn't feeling well and went to lie down. He complained of chest pains, so we immediately drove him to a nearby doctor's office. Sadly he died there. The board chairmanship was passed to Bill Rubey, who in 1968 had been appointed director of the newly established Lunar Science Institute that was funded by my office. All the attendees agreed to continue the meeting, and the Academy and Board quickly issued a report strongly supporting the remaining missions. However, we had lost a strong and influential voice as we continued our battle to assure that the missions would include critical science experiments and return the best science results possible. The number of flights that would follow Apollo-11 was still uncertain.

The success of Apollo-11 did not signal a lessening of effort as we prepared for the flights that followed. On the contrary, my workload increased as the science to be conducted on the remaining flights was much more extensive and complicated than what had been done on Apollo-11. Several detailees from Jet Propulsion Laboratory were added to my staff. Our hope to follow the first Apollo missions with lunar bases and wide-ranging exploration now seemed remote, but important work still lay ahead to make each succeeding mission scientifically productive. The final three landings, designated "J" missions, would allow us to conduct the most extensive exploration and carry the largest science payloads; thus they required the most preparation.

As one of the tens of thousands who had been hired to work on the Apollo program, and consumed with the desire to assure its success, I often worked seven days a week. Meetings were held at NASA centers all over the country as well as at many contractor facilities. Most weeks I had to take at least one trip. My two small sons needed a full-time father's attention, and the constant travel began to take a toll on my marriage.

I attempted to compensate for my absences in many ways. When at home I coached the little league softball team and was a Cub Scout den father as well as the leader for Cub Pack 782, a total of over 50 boys. I took Tommy and Bruce camping, to Washington Senator baseball games and my Navy squadron's reviews at Andrews Air Force Base. But disagreements with Pat grew as to how to maintain the type of family life she desired. A trivial argument became a long-lasting conflict. We had, as the saying goes, become incompatible, and I didn't want Tommy and Bruce living with incompatible parents. Long before the final Apollo mission splashed down we separated. I moved to an efficiency apartment in DC and began divorce proceedings.

Astronaut Training

I was able to participate in a number of the field geology training trips, and all were memorable. One of the last exercises with the Apollo-17 crew will give a flavor of how they were conducted. The site selected to review procedures for each of the final three Apollo missions was the island of Hawaii. Despite the prevailing view that most lunar features were the result of impact processes, early in their training all the astronauts had visited Hawaii to study the wealth of lunar-like features created by the many active or semi-active volcanoes. Perhaps a lunar landing site might include an extinct volcano, as it wasn't certain that all the craters were caused by impacts.

Simulations in Hawaii were designed like final exams. Several locations on the island represented geological problems and features similar to those that had been mapped at their landing site based on Lunar Orbiter photos. For the Apollo-17 crew, the first four days were spent visiting these geologic features. Dallas Peck, a noted USGS volcanologist who had spent a number of years on Hawaii studying the island's geology, acted as coordinator and principal lecturer. The final three days were spent at Kahuku, Hualalai, and the volcanic ash wastelands at the crest of Mauna Kea. The traverses laid out for this exercise reproduced, as closely as possible, those that Cernan and Schmitt would follow at their designated lunar landing site, the Taurus-Littrow valley. USGS staff had prepared the routes with sampling and description stations replicating the timeline already carefully plotted by the Field Geology Science Team for the actual mission.

Replicas of all the equipment that the crew would use on the Moon, with the exception of the ALSEP, were transported to the top of the crater, including a version of the Lunar Roving Vehicle. The flight version of the LRV could not be used in terrestrial simulations because it was designed to operate in lunar gravity; it would have collapsed under the astronauts' Earth weight. Putty Mills, the USGS LRV guru, had modified a local jeep as a simulated LRV. He had removed most of the jeep's body so that Cernan and Schmitt were sitting on the frame on open seats that allowed them to climb on and off without difficulty. He had also added racks for their tools and sample bags similar to those on the lunar LRV, and a mount for a communication antenna.

Cernan and Schmitt wore street clothes for these simulations; it would have been too costly and time-consuming to use pressure suits this far from Houston. But to add some mission reality they wore backpacks similar in size to the Portable Life Support Systems they would carry on the Moon. The backpacks had battery power for voice communication back to our simulated Mission Control Science Support Room (SSR) located on the mountaintop out of sight of the traverses. Those manning it pretended they were at the SSR in Mission Control.

Bill Muehlberger, the Field Geology Team leader for Apollo missions 16 and 17, was in charge of this trip. He brought with him several members of his team, including George Ulrich, Gerry Schaber and Dale Jackson. Scientist-astronaut Bob Parker was also on hand. He had been designated mission scientist and would be the prime CapCom when Cernan and Schmitt were on the lunar surface during EVA periods. Muehlberger and his team manned the SSR, plotting the astronauts' progress as they drove around the volcano's summit and communicated through Parker as they would during the actual mission.

Those of us not directly involved in the SSR simulation followed Cernan and Schmitt from a distance as they drove from station to station, making note of how everything fit together, or didn't, as the case might be. The Field Geology Team, through trial and error on earlier missions, had devised procedures to assist the astronauts when on the Moon if something unexpected happened, or to respond to questions they might have. These procedures were also practiced, relaying questions and suggestions through Parker.

At the end of the exercise Muehlberger and his team retraced the traverses with Cernan and Schmitt, reviewing how they interpreted the astronauts' voice reports. Muehlberger then suggested ways for his team to better interpret what the crew actually saw and the actions they performed such as sampling and photography.

With the first scientist-astronaut geologist in the crew and a highly motivated and well trained commander, we didn't expect there would be much need to provide real-time support but, as with all things NASA, we were prepared. All in all, this simulation was about as good as we could get in obtaining a high fidelity rehearsal before the real mission was underway. One week of intensive, almost uninterrupted training helped both the crew and the Field Geology Team. Apollo-17 would be the last mission, and Muehlberger was determined that it would be the best. In just five months it would be the real thing.

A final reward for our efforts had become a tradition. On the last night of the trip a dinner was held at the lovely Teshima Japanese restaurant high on a hill overlooking the ocean. Mrs. Teshima provided a royal welcome and a special menu. It was a night of story telling, practical jokes and reminiscing that I am sure all who attended will never forget.

Broadcasting Apollo Missions on Voice of America

In addition to my NASA duties VOA asked me to continue to support its worldwide English and Spanish language broadcasts, the latter using my still serviceable Colombian Spanish. For each mission, after coordinating the pre-launch science press briefing at a Cocoa Beach hotel and attending a few "blast-off" parties hosted by private companies the night before, I sat with Rhett Turner on press row to describe the launches and what the astronauts would be doing.

Thus, on December 6, 1972, Apollo-17, the last of the Apollo lunar flights, was on launch pad 39A. My good friend Jack Schmitt, with Gene Cernan and Ron Evans were strapped in on board the command module. As we had done for the last two flights, Rhett and I were prepared to broadcast the launch for the VOA Worldwide English Service from our press site about three miles west of the pad. To our right, along the raised, curved berm were the large air-conditioned broadcast booths of CBS, ABC and NBC, along with booths of other domestic and foreign companies. Through our binoculars we could see Walter Cronkite, Jules Bergman and the other TV commentators looking out of their large picture windows at the brightly illuminated Saturn V. Voice of America was a bare-bones operation. Last in line on the berm, we sat in the open on two folding chairs beside a folding card table, swatting mosquitoes. Behind us, in what was once a small two-wheeled camper trailer, was the engineer with all the necessary electronic equipment.

Apollo-17 was scheduled for launch at 9:53 PM EST, the first and only night launch for Apollo. So, in addition to our excitement at being present at this last launch and all that it meant, we were looking forward to seeing and "feeling" the giant Saturn roar off the pad, a sight at whose visual impact we could only guess. The world's largest firecracker would light up the night sky. As we sat watching the brilliantly lit launch tower and rocket, the clouds that had partially obscured the sky slowly dissipated. It had been predicted that in a clear sky viewers from as far away as 500 miles might see the rocket as it streaked away to the east.

Rhett, the best space reporter I knew, had done his usual impeccable homework for the launch. He carried the radio audience along as the countdown proceeded, bringing me in as needed to provide some special insight. At times during our rather informal script Rhett would signal the engineer to play a previously recorded interview with Jack, Gene or Ron, or some other pertinent clip he had made that would provide interesting background about the mission. In front of us, between our position and the pad, was the brightly lit large digital clock, counting down the minutes and seconds. From time to time Chuck Hollingshead, the voice of Kennedy launch control, would interrupt our coverage with comments broadcast over the PA system.

Everything was proceeding normally until T-minus 30 seconds when, without any warning, a hold was announced. No immediate reason was given, and we were left to speculate with the rest of the commentators. For the next 20 minutes we tried to make educated guesses as to what the problem was, meanwhile expressing our hope that it was something minor and the countdown would soon resume. Finally Hollingshead came on the PA speaker and explained that the third stage fuel tanks had not pressurized on schedule; when manual pressurization was attempted it was too late in the countdown, and the automatic sequencer had shut the launch down. It didn't sound serious, but it wasn't clear when the count would start again. We assumed that perhaps in one hour the problem would be fixed and the countdown would resume at T-minus 22 minutes, the announced recycled starting point.

We were wrong. For the next two hours Rhett and I filled the airways with impromptu discussions of Apollo-17 science and whatever other subjects we could think of. From time to time VOA would break away to provide news of the world and return to us, still sitting under the stars waiting for an announcement that the countdown would resume. Listening to those VOA tapes many years later is a real trip down memory lane. The tapes recorded late-breaking news, including an item that former President Harry Truman, age 88, was in critical condition in a Missouri hospital. Recently I discovered that the National Archives had saved over 700 VOA Apollo tapes, none of which were heard in the US as VOA wasn't allowed to compete with domestic radio and TV stations. With Rhett's help Rob Godwin and Apogee Books is reissuing the broadcasts along with never before seen videos of interviews with the crews. They will be available to history buffs.

Finally the count resumed at T-minus 22 minutes and held again at the planned point of T-minus eight minutes to check that the pressurization problem had been resolved. At 12:33 AM on December 7, 1972, Apollo-17 was launched, an historic date, one that will be remembered along with another December 7th 31years earlier. The liftoff was every bit as spectacular as we had hoped, lighting the night sky for miles around and pounding our bodies with the powerful, low-frequency reverberations that only a Saturn V launch produced. If you have witnessed a shuttle launch, multiply the effect by two. The crew of Apollo-17 was on its way. To top it off, we had survived two hours of unscheduled air time! We packed our notes and left for Houston.

Apollo Epilogue

One of the most frequently asked questions of the Apollo program is what did we learn. It would often be followed by other unanswerable questions. Was the program worth the cost? Wouldn't the nation have been better off spending the billions of dollars on some other national program such as health care or subsidized housing?

Addressing the unanswerable questions first from the perspective of science, it is very difficult to calculate the part of Apollo's cost that funded science experiments. This is so because through the years the many components that made up Apollo science were carried in different parts of NASA's budget. I made an estimate of $350 million based on summing the cost of astronaut science training; the cost of each experiment, its integration, data reduction and analysis; the cost of the quarantine facility for the returned samples and their analysis by hundreds of scientists; and other costs that included NASA salaries. Using this number, the science portion of Apollo was about 1.5 percent of the total $25 billion spent. Science was not a major driver in the overall cost of Apollo.

On the plus side are the prestige value of being the premier spacefaring nation, the excitement of visiting a new world and the knowledge gained about the Moon and Earth. Much of the advanced technology that was needed to assure Apollo's success was passed on and used by US industry. Considering the above, was it money well spent or would it have been better spent on improving social programs? I believe it was money well spent.

"What did we learn?" is an important question but difficult to quantify. Lunar scientists continue to examine material brought back during the Apollo missions, and fresh results keep coming in from other lunar programs. Information from missions undertaken after Apollo continually adds to our knowledge and clarifies or extends the Apollo results. It would be impossible to summarize here all we have learned about the Moon and in turn how it has helped us understand more about Earth. I will leave it up to the curious and industrious to do the research.

One afternoon, while completing my final Apollo reports, I was walking between my office at L'Enfant Plaza and NASA offices at 600 Independence Avenue. I met some of my former colleagues who had recently been assigned to manned space flight programs that were just beginning to be defined. We talked briefly about the uncertainties surrounding these programs (none had been officially blessed as Apollo's successors except for short-term Skylab and Apollo-Soyuz programs) and about where I might find a new job. As we discussed NASA's future I was struck by their lack of enthusiasm for their new assignments. It seemed as if suddenly, almost overnight, this marvelous can-do agency had grown old and lost its way. Grey heads were beginning to predominate in all the offices. Instead of looking ahead to an exciting future, everyone seemed to be scrambling to hang on and find a place to roost. I decided at that moment it was time to leave NASA and find a new outlet for my energies.

Following the lead of several former colleagues who had already left NASA, I sent an application to the National Science Foundation. I was hired immediately by an office that was undertaking research on a wide spectrum of new technologies. Thus began a new career, but never again would I experience the excitement and sense of achievement that came with being a small part of the Apollo program.

National Science Foundation -1973 to 1975

My new job at the National Science Foundation (NSF) began as a staff position but held the promise of managing research programs to develop new energy technologies. Although this would be a different direction in my government career, my new boss was Al Eggers, a former NASA manager. I didn't know Al while he was at NASA as he was working at the Ames Research Center outside San Francisco. Besides participating in the early planning for Apollo, one of his NASA contributions had been the design of "lifting bodies" that allow spacecraft to survive high-speed reentry into the Earth's atmosphere. In his new position he had convinced NSF management and Congress to start a program of applied research as opposed to NSF's traditional role of supporting basic research. In 1971 a new NSF Directorate was formed under his management, eventually given the title Research Applied to National Needs (RANN). The establishment of this new NSF directorate coincided with a speech made by President Nixon in June calling for a program to develop supplies of "clean" energy. US oil production peaked in 1970, and it was clear that the nation would need to develop and utilize other energy sources if the economy was to continue to grow.

For his new Directorate Al chose as his deputy Dick Green, who had been part of the NASA team that managed the early development of Apollo science experiments. Also on Al's staff was Ed Davin, a former member of Will Foster's office. Working with former NASA colleagues was something of a homecoming. Adding to the feeling, NSF's office on G Street was one block away from where I first began working at NASA in 1963 and only a half block from the White House compound.

After a few months adjusting to NSF culture, very different from that of NASA, I was put in charge of the Advanced Energy Research and Technology Division. My responsibilities included managing programs to develop technology for solar and geothermal energy applications, energy conversion, transmission and resource recovery. Some of this work would allow me to apply skills learned ten years earlier at Mobil Oil, such as drilling wells to recover geothermal heat, one of my new responsibilities.

My new staff was very small, just four professionals and a secretary. Considering that my 1973 budget was a little less than $4 million, it was a manageable workload. Importantly, reflecting administration and Congressional concerns, it was more than three times the budget provided two years before I arrived. We carefully evaluated hundreds of proposals and selected 50 to receive grants in seven categories, each averaging less than $100,000. These were big dollars for traditional NSF grants but small potatoes compared to what I had managed at NASA. The good news was that for many years before RANN was established a nascent solar energy research community had existed that jumped at the chance to receive government funds to continue its research. Thus it was not difficult to find established researchers at universities, not-for-profit organizations and in the private sector who had worthwhile research projects.

To be selected for a RANN grant, a proposal had to show that it would be possible to take the research to a demonstration in a short time and preferably include an end-use partner or a private sector company that would also contribute resources. This requirement was completely new to NSF and provoked considerable grumbling from the established research directorates. We ignored this criticism as the RANN charter specifically endorsed such cooperation and it meant that small amounts of NSF funds went much farther toward accomplishing useful objectives.

Competition among government agencies to become involved in energy research grew in 1973. Other agencies, including NASA, attempted to capitalize on the growing interest in renewable technologies. Competing for scarce dollars in budgets sent to Congress by the president is an old game in Washington. We were positioned to be a leader in developing alternative energy sources and convinced the Office of

Management and Budget that RANN would be the best agency to lead government-sponsored research efforts to develop and demonstrate renewable energy technology.

Divorce and New Marriage

At the beginning of March my divorce became final, a sad and difficult ending to what once had been a promising future. Pat was given custody of Tommy and Bruce while I provided child support, health care, funds for their future college education and other benefits until each reached the age of 22. Pat retained the house and furnishings and I would start over again.

During the separation before the divorce became final, on one of my frequent trips to Houston I met a pretty Irish lass. She sat across the aisle on an Eastern Air Lines flight originating in Boston that I boarded at Dulles. She had a small red cooler under her seat; out of curiosity I asked and learned it contained live lobsters. As we were deplaning I asked if anyone was meeting her and volunteered to drive her home, which was only a little out of the way on the route to my hotel in Clear Lake.

From that chance encounter a friendship began with Ann Snyder, nee Keane. On future trips to MSC I called and we met for dinner or another social event. At one of these I met her friends Charles and Melba Troub, an encounter that resulted in a hilarious adventure in Mexico. Charlie owned a radio station in McAllen, Texas. He was a wheeler-dealer, always promoting one thing or another. He had been instrumental in fostering a "sister city" program with Reynosa, the Mexican town just across the Rio Grande. He knew all the important people on both sides of the river, including the governor of the Mexican state of Tamaulipas.

For the big unveiling when the samples returned by Armstrong and Aldrin were released from quarantine for study and display, I took one to a Congressional hearing and brought Tommy and Bruce along to participate. It was one of the larger pieces, a sparkling breccia mounted in a handsome triangular-shaped plexiglas case. One could view the rock from all angles. Even to an untrained eye it was impressive. Charlie, remembering the story of the Congressional hearing, asked if I could bring the rock to McAllen for the upcoming Tamaulipas state fair. I told him I couldn't bring the actual Moon rock but I could bring a replica. For just such events the Lunar Receiving Lab had made several very accurate reproductions, all mounted in identical cases. He hesitated, then said, "Bring it down."

With the support of my boss, Lee Scherer, I received permission to show the replica at the fair. One of the reasons my trip was approved was that I could discuss the sample in Spanish and describe the mission as I had on Voice of America. I picked up the replica in Houston and drove to the Troub home in McAllen. Charlie looked at the replica and said, "I told the governor that you were bringing an actual Moon rock. Play along with me and don't say it's a replica." With some misgivings I agreed.

The next morning we drove across the bridge to Reynosa where we were met by a police escort. With sirens wailing we drove to the fair grounds. I met the governor, took the replica out of its carrying case, and described the rock. Then, with a police guard at our sides, we walked to the special tent that had been set up to display the "Moon rock." I placed the replica on a red pillow on a small table illuminated by spotlights. Armed guards were stationed at each side of the table, and dignitaries and fair attendees were permitted to walk slowly past the display. Charlie's hoax was working. The next morning the Reynosa newspaper had a headline and picture on the front page showing the "Luna Roca"!

However, there is more to the story. At the close of the fair's first day I went back, put the rock back in its carrying case and carried it to the police station. There I watched as it was locked in a large safe. The next morning Charlie and I went to the police station to retrieve the rock for its final day of display. The police chief opened the safe and handed me the case. With a quizzical look he asked, "Are you sure this is a real Moon rock?" I saw nothing unusual in his question and said yes. We put the replica back on its pillow for the last day. Then I picked it up and drove back to Houston. When the lab staff opened the carrying case and looked at the replica, still secured in its plexiglas case, they noted a small scratch on one side that had not been there when I first picked it up.

Apparently the Mexican police chief couldn't resist temptation. That night he had opened the safe, carefully took apart the plexiglas case, and scratched the replica with a knife. The fake rock had been made by carefully forming the shape of the real rock in plaster of Paris. Then it was covered with material resembling the coating that had accumulated over the billions of years it had been exposed to the cold, hard vacuum of space. The tiny scratch revealed the white plaster of Paris. Did the Reynosa police chief believe it was a fake, or did he think a Moon rock's inside really looked like that? I'll never know, and probably he won't either. However, Charlie finally "fessed up" to his hoax and the mayor of Reynosa sent a very nice letter to Lee Scherer thanking him for lending the replica for the fair.

Slowly Ann and I became more than friends. She had just gone through a difficult divorce, so we had experienced similar woes. Born in Ireland, she had been educated and trained in England and was a registered nurse working at a Houston allergy clinic. She was the oldest of four children, two boys and two girls. Her sister Maeve, also a nurse, had recently emigrated to the US and was working at a hospital in Boston. Ann's family had an impressive medical history. Her grandfather had been the doctor for the isolated Aran Islands off Ireland's northwest coast, and two uncles were also doctors. Ann was returning from a visit with her sister when we first met on the Eastern Airlines flight.

As our relationship grew, we began to believe we had so much in common that we could have a successful marriage. On the last day of March, 1973, we were married by a justice of the peace in Sugarland, Texas. A wonderful reception followed, hosted and attended by her many Houston friends.

As part of her divorce settlement Ann had received a small house on Houston's south side, within walking distance of many well known Houston medical facilities. It was put up for sale and quickly sold. We hired a moving company and packed up her household effects. We put a few special items in her red 1970 Oldsmobile convertible, including her Sealyham Shaun and cat Christmas, and moved to Rockville, Maryland. I wanted to be near Tommy and Bruce to participate as much as possible in their lives. We went house hunting in a neighborhood close to my former home where Pat and the boys still lived. We found one we could afford, a great home that we improved through the years and in which we entertained and Ann served many wonderful dinners for friends and guests.

Tommy was attending Earle B. Wood Middle School and would soon be a freshman at Robert E. Perry High School a half block from home. Bruce was in Aspen Hill Elementary, the same school that Tommy had attended. In this regard not much had changed in their lives, same house, same friends. But there was no dad. It was a difficult time for me and I believe for them. Tommy was a good athlete and had attended summer basketball camps. He was on the Wood basketball team and would make the Perry varsity. I attended every game I could and kept a scrapbook of newspaper clippings of his games and other activities. Bruce was a hard worker. He had a morning newspaper route for several years and saved his money. I am sure he could also have been a good athlete, but I wasn't there for him every day during his critical pre-teen years as I had been for Tommy. The encouragement and sharing of simple things like catching a baseball or football in the back yard were missing.

We had Tommy and Bruce for an occasional dinner or lunch, and Ann cooked a great meal while staying quietly in the background. In the summer of 1974 I rented a motor home (I sold the Winnebago in 1973) and the boys and I drove to Colorado and back, stopping to camp, fish and canoe wherever the mood dictated. I showed them my thesis area and other favorite places in Colorado. But it wasn't quite the same as our earlier motor home vacations. I think they enjoyed our time together but am not sure. Skipping ahead, both went to the University of Maryland. Tommy graduated. Bruce didn't complete his degree, a result I am sure of the lack of continuous guidance on my part.

New Neighbors

Our house, five years old when we bought it, was originally one of the builder's model homes. Nextdoor was another former model home and one of the first occupied in our development. It was owned by an elderly couple whom we came to know well over the next 20 years, Rudolf and Ariadne Loewenthal.

Rudolf, a German Jew, fled Germany in 1936 and ended up in Shanghai. He told me that at the time there was a sizable native Jewish population in China plus many refugees. He was embraced by that community and eventually taught at the Catholic University in Peking. He told me one of his pupils was Zhou En Lai, first premier and foreign minister of the Communist People's Republic of China.

At the end of WW II Rudolf came to the US and joined the faculty at Cornell University. He was a prolific writer, with many works published before and after he left China. At some point, I believe in the 1950s, he met and married Ariadne Lukjanow. When we met them Rudolf was still busy collating and publishing what he called bio-bibliographies of famous Jews who had lived in China.

Ariadne's history is even more fascinating and tragic. Her father, Wladimir Lukjanow, came from a prominent Russian family. His mother's family, leaders of the Ussuri Cossacks, once lived just west of Moscow but during the reign of Ivan the Terrible emigrated to Siberia. Eventually they lived in Vladivostok, where Wladimir was born. Apparently a very bright young man, he attended a university in Tomsk and then was sent to the University of Grenoble in France where he received PhDs in civil engineering and chemistry. While in France he took up painting. After receiving his degrees he returned to Vladivostok with a divorcee, her son and his painting teacher, much to his family's chagrin. He married the divorcee, but she soon died of tuberculosis.

Before the Soviet revolution the Lukjanow family was very wealthy, owning a glass factory and other businesses. After the revolution they were forced to give up their factory, and Ariadne's father began to work for the communist government in Siberia. After his first wife's death he married Ariadne's mother and Ariadne Wladimirofna was born in 1921 in Vladivostok. In 1922 Wladimir moved his family to the Crimea, where he was appointed head of the Southern Russia Ukraine-Crimea region and developed the region's first five-year plan.

But he was not a happy communist camper. When Germany invaded the Crimea in WW II he immediately took advantage of the turmoil, fled with his family to Germany, and joined Breslau University. Toward the end of the war, as the Soviets were approaching, they escaped to Bremen in the British zone, where they lived under the sponsorship of Allied intelligence. Obviously, someone with Wladimir's background was very much sought after. He came with his family to the US in 1949, eventually settling near Chicago where Ariadne met Rudolf. After they moved to Washington Wladimir began painting again, recording remembered scenes of his life in Russia as well as local landscapes and portraits of historical figures.

After marrying Rudolf Ariadne worked in the field of machine translation. At one point she had a contract with NASA. One day at work she learned her father had been badly injured in a traffic accident. She drove to the scene and accompanied her father in the ambulance. On the way to the hospital it was involved in a serious accident. Ariadne suffered life-threatening back injuries. When we moved nextdoor and met her, she was taking heavy doses of pain medication and had great difficulty moving about. They had no close friends in the area, thus through the years we alternated hosting dinners, and in this way I slowly accumulated these stories.

One morning Ariadne committed suicide, hanging herself in the basement. Rudolf became a lost soul living by himself. As he had no relatives to turn to, we helped him as best we could. He came to dinner often, and each time he brought an unframed painting by Ariadne's father. I frequently brought meals to him that Ann had prepared, and she did some shopping for him. One morning I knocked on the door, had no response, and went in. I searched the house, finally going down to the basement. There I found Rudolf standing on a chair on tiptoe, a length of clothes line wrapped around a beam and tied tightly around his neck. His face was turning blue as I cut him down. He survived, and with the help of Montgomery County services we placed him in an assisted living facility.

With Rudolf's permission we put his home up for sale and had an estate sale of the furnishings. I donated his unique library collection to George Washington University. Almost everything sold except for a few pieces that I felt didn't receive good bids. We bought them at the appraised price. A beautiful Chinese screen with delicate scenes of inlaid mother-of-pearl and ivory that Rudolf had received as a gift from the Catholic University in China he gave to us. With the proceeds from the house and estate sale we were able to move Rudolf to a very nice assisted living facility near Rye, New York. An old China acquaintance, whom he had named in his will as his only beneficiary and lived near Rye, now became Rudolf's guardian. He died in 1986 after suffering a stroke. We still have all the pieces we bought, the screen and Ariadne's father's paintings, now nicely framed. Perhaps one day they will be passed on, along with this tragic story.

Energy Research Programs

After Ann and I were married my work at NSF continued at an increasing tempo. In June 1973 NSF was designated the government's lead agency to manage renewable energy research and development. Solar energy can be utilized in many different ways: low temperature systems to heat and cool buildings, high temperature systems for many applications using concentrator arrays, direct conversion of sunlight to electric power using photovoltaic cells, wind turbines and power conversion using hot and cold ocean water to generate a low temperature "steam" cycle. With such a wide variety of potential applications we were inundated with proposals. Most were legitimate, a few came from charlatans, and some suggested they could develop perpetual motion devices using solar energy. I needed more staff, and my office grew.

Also in June President Nixon asked Congress to initiate a five-year, $10 billion program to develop energy technology. Overall management of this program was given to the Atomic Energy Commission (AEC) that was responsible for developing peaceful uses of nuclear energy. A multi-agency team was put together with the solar panel chaired by Eggers. This marked the beginning of what became a well publicized dispute between Congress and the AEC over how technologies other than nuclear and fossil energy could contribute to the nation's energy future.

Before AEC could submit a report on how the program would be structured, the OPEC oil embargo was put in place in October 1973 and the "ball game" suddenly changed. Long lines at gas stations, pictures of children sitting bundled up in cold classrooms and other effects of the embargo raised the cry for energy independence. I immediately established and chaired a panel of 17 government agencies to identify and develop applications for solar energy.

With a small staff and programs growing, I worked six or seven days a week. On Sundays Ann came with me to the office. While I cleared my in box and got organized for the week ahead she read our two Sunday papers, marking articles of interest for me. Then we had lunch at a local restaurant and returned home. This became a standard and necessary routine.

Budgets for my office grew rapidly. At one point Al Eggers, Dick Green and I testified before a House subcommittee asking that we not receive any more funding for solar energy R&D than already appropriated because it would be difficult to spend it wisely. This may have been the only time in history that Congress heard such a plea! But the committee disregarded it and voted to increase our funding, perhaps a testimonial to their belief that we were actually accomplishing important results with current funding and could do better with more. We spent the increase building solar demonstration projects all over the country, some attached to schools to provide heat and hot water.

Project Independence and Bilateral Energy Agreements

As the oil embargo continued into 1974 President Nixon decided to undertake a major study, Project Independence. The name said it all; in the future the nation must not depend on energy supplied from sources beyond our immediate control. With this announcement our workload increased significantly. I chaired the Solar Task Force and co-chaired the Geothermal Task Force. Our findings would be part of the final report. For the remainder of 1974, until the report was issued, we were involved in numerous meetings with both task force members and the combined Project Independence team.

The Nixon administration also negotiated several bilateral energy agreements with other nations to encourage an exchange of ideas and technology development. NSF became a partner with the Soviet Union in two agreements covering solar and geothermal energy. In our first exchange on geothermal energy a team of Soviet specialists came to the US for tours of our research facilities including the Geysers geothermal plant in California. Geysers was, and probably still is, the largest application of geothermal power in the US.

Our reciprocal visit to the Soviet Union ended in disarray. The most important Soviet projects were located on the Kamchatka peninsula. At the last minute, as the team prepared to fly from Moscow to Petropavlovsk, the visit was canceled. The reason given was a problem at the Petropavlovsk airport, but we suspected the Soviets believed we intended to spy on their military bases on Kamchatka. In 1974, when overhead surveillance did not provide quite the detail it does today, Kamchatka was declared off-limits to foreigners.

Our next exchange involved solar energy. Before starting our visit I took a crash course in Russian and attended night school for three months. I didn't become proficient but picked up enough vocabulary and grammar to speak some simple sentences, greetings and other useful dialogue. My team of eleven included an NSF staffer, John Thomas, who I believe lived in the Soviet Union in the 1930s when his father, a Ford engineer, went to Gorky to help set up an automobile assembly plant. John was completely fluent, and his presence turned out to be very useful.

In September 1974 we flew to Moscow to begin the visit which was scheduled to include seven cities where we were told important solar energy research was taking place. My team of experts was comprised of government and university researchers. The opening meeting in Moscow was contentious. The two delegations sat on opposite sides of a long table covered with a red cloth. Flags of the two countries and bottled water were placed down the center. At the outset we were confronted with last-minute changes to the agreed itinerary. I was less than diplomatic because of the problems encountered by our geothermal team during its previous visit. I turned to John Thomas and said, "Tell them we are going to cancel our visit and return to the US." The US Embassy representative present was shocked and urged me to reconsider.

Apparently the Soviets weren't prepared for this reaction and quickly said they would reexamine the itinerary and try to return it to the original agreement. I accepted this and with some misgivings we departed for Tashkent, our first destination. We were assigned two Soviet interpreters, a woman and a man, the latter a former paratrooper, both very friendly and accommodating. However, our misgivings were justified and we never were allowed to visit all the laboratories listed on the original itinerary. To give you a feel for the "atmosphere" in which we operated, when I wanted to discuss a subject privately with any member of the team we walked outside whatever building we were in, as we had been warned that conversations inside buildings were monitored. John would fill me in on the small talk he overheard that wasn't translated by our interpreters that indicated how the exchange was being evaluated.

At Tashkent we visited projects at several research facilities, but an important project could not be seen because, we were told, the bridge connecting it to the access road was under repair. Get the picture? On the final morning in Tashkent, as we prepared to board the bus to the airport and fly to our next destination, I counted noses and came up with only ten. I found our missing colleague still in bed suffering a heart attack.

He was our representative from the Aerospace Corporation. Although he was a recognized solar researcher, he told me that he was privy to highly classified information. The Aerospace Corporation was an Air

Force "think tank" that conducted many classified studies. He told me he could not remain by himself as he feared the Soviets would subject him to interrogation while sedated. My only choice was to leave another member of our team, Vic Bremenkamp from Associated Universities, NSF's contractor who handled all the trip logistics and was not a solar researcher. Vic stayed with our Aerospace member until he recuperated and rejoined us. The rest of us got on the bus to the airport and flew off to Ashkhabad.

The trip proceeded for the next week with a stop in Yerevan, Armenia, and a continuing rash of health problems as team members succumbed to one malady or another. The Embassy had given me a small medical kit that included pills for various ailments; apparently it was aware of problems other delegations had encountered. Each morning I would hold a "sick call" and dispense those medications that fit particular problems. One night only one other member and I were able to attend a big dinner given in our honor; the rest were in their hotel rooms, most suffering from stomach ailments. My stomach, used to life in the jungle, seemed to be immune.

We found, instead of technical discussions, the Soviets more inclined to show their hospitality with large, lengthy luncheons and dinners taking up much of each day, all accompanied by frequent vodka toasts. At one luncheon I noticed that between toasts my Soviet counterpart was surreptitiously pouring his vodka into a potted plant. I made a joke of his action in one of my toasts, much to his embarrassment. In addition to the everyday official luncheons and dinners we were treated to two evening events, a ballet and circus, the latter including many very funny acts. Our hosts seemed very proud that they could show us these aspects of Soviet culture and we, of course, said how much they were enjoyed.

The Soviets seemed convinced that I was a US intelligence agent. To add to this perception, for my amusement, during meetings while discussions were being translated, I pretended not to understand any Russian, which was mostly true. But when it was my turn to respond through their interpreter I had picked up enough of the discussion so that it appeared I had understood everything the Soviets were saying before it was translated to English. I was followed everywhere by little men taking my picture. I wonder if the former KGB had a dossier on me.

When we reached Baku, the last stop before returning to Moscow, another team member came down with a very serious problem, bleeding ulcers. He had had them many years before, and one of the remedies was to stop drinking alcoholic beverages. He thought that for a short time on this trip, to be courteous, he could drink again. Wrong idea; he called me to his room. What to do? I asked our Soviet host if I could call the US embassy in Moscow for instructions, and he led me to the hotel's basement. There, in a large room filled with desks, most empty and with a single phone on each, he gave me a red phone with a direct line to the embassy. I was advised to take my "patient" to a local hospital, a military facility. There they froze his stomach and kept a woman attendant with him at all times. A few days after we returned to Moscow he rejoined us in fairly good shape. The Tashkent heart attack victim and Vic had rejoined us in Tbilisi, a scheduled stop, before we arrived in Baku.

To make a long story short, there were many other frustrations including another canceled visit to a well known research institute. Want to guess the reason? The airport was undergoing repairs and only small planes were allowed to land.

Based on suggestions made by colleagues who had visited the Soviet Union, I included in my luggage some easily carried gifts to give to our hosts at each stop. They consisted of records of the latest hit songs that I was told their children would appreciate as they were difficult to buy in the Soviet Union and small albums of US postage stamps that I selected from duplicates in my collection that I had begun after college. All seemed to be appreciated and after being presented reduced the weight of my bags. We were permitted to shop at Berioshka stores reserved for those with dollars or hard currency. I bought a few trinkets, an atlas of Soviet Union maps and a few books to remember my visit.

There was a final problem. Two days before we were scheduled to leave Moscow one of the official hosts who had traveled with us died. To show our respect I asked Vic to stay and attend the funeral that was to take place the day after we left. He waited at the hotel to be taken to the ceremony as promised, and waited and waited. No one ever came. In the two days of waiting his visa expired. When he arrived at the airport for his rescheduled departure he was not permitted to board his flight. He was held up for another day before the embassy came to his rescue.

When we departed and our plane's wheels left the runway at Sheremetyevo airport, all the Americans on board applauded. We were no longer being watched and followed day and night. One great reward for me was that on the way back home I met Ann in England. We had a wonderful vacation, staying first at a former Elizabethan manor and fishing lodge, Gravetye, that had been converted to an eight-room hotel. There were beautiful gardens and not one clock anywhere on the premises. When we joined the few guests in the formal sitting room the second night of our stay, without asking the waiter brought us the same cocktails we had ordered the night before. Gravetye was listed as a five-star hotel and lived up to the rating. We enjoyed good food (food we were served in the Soviet Union often failed a taste test) and interesting sightseeing, including Winston Churchill's home, all free from constant surveillance.

From Gravetye near East Grinstead in southern England we drove north in our rented car to Wales, where Ann had attended St. Clare's boarding school. We dealt with street signs in the impossible Welsh script but eventually found Porthcawl. We stayed in a quaint bed and breakfast and located Ann's old school building, but it was no longer in use as a school. Then we drove to Holyhead and late at night took the ferry across an angry Irish Sea. From Dublin we drove north to Dundalk, where I met for the first time some of Ann's family, including her widowed mother Gertie. We stayed for two days at Ann's old home, warmed in the chilly evenings by burning "turf" (peat) in the fireplaces. At the local pub we met the parish priest who thanked us for the rounds of spirits we bought. I assume I passed inspection as Ann's new husband, as everyone was very friendly. Then we flew home to make my official report on the visit to the Soviet Union.

The Soviet reciprocal solar visit to the US, scheduled after we returned from Ireland, went well. The itinerary proceeded as planned, and the Russians toured several universities and companies involved in solar research. Their visit ended with two parties, the first at our home in Maryland attended by all my office colleagues. Our Soviet guests seemed surprised at the size of our home (typical of our neighborhood) and thought it was a government-owned house made available for the party. They insisted on a tour. Perhaps they were looking for "bugs" and hidden cameras that were probably in all Soviet buildings. They went through every room, upstairs and down, looking in closets and wherever else they wanted. Only when I showed them my work bench, covered with sawdust and tools, did they finally believe it was a private residence.

The next day Lloyd Herwig, a leading solar energy researcher on my staff who was on the trip to the Soviet Union, hosted a cookout at his home in Virginia. The Soviet delegation also wanted to tour the Herwigs' house. This turned out to be a little embarrassing for Donna Herwig. Their teenage children's rooms in typical disarray convinced her visitors as she showed them around that they were in a private home, not one supplied by the government for the occasion. The "cold war" spawned many strange beliefs on both sides, but the Soviet visitors' firsthand exposure to the US, our open society and life style, seemed to surprise them and perhaps made a lasting impression.

Despite the cordial ending to the exchange I recommended that the bilaterals be canceled. It was clear that there was little to be gained for our side, not to mention all the trumped up problems we encountered. Solar energy research in the Soviet Union had a long history, but we were shown only old applications and projects. A few appeared to be dusted off and started again for our visit. In the few private discussions we were able to have with our Soviet counterparts, when their "handlers" were not listening, we were told they were not hiding anything but were not receiving sufficient funds to carry out new research and hoped the bilateral exchange would lead to receiving additional funding. The bilaterals were allowed to expire.

A final footnote: some months after their visit, the Soviet team leader called me. His wife was suffering from a medical problem; could I send some medication she needed that was not available in Moscow? I asked the State Department what I should do. I never received a reply, so I let the matter drop and didn't call back. We learned a short time later that his wife had died. Although a relatively young man, he also died a few years after our meetings. All the Soviet professionals we met were heavy smokers, and perhaps his habit claimed him.

The final Project Independence report was issued in December 1974 by the newly formed Federal Energy Administration. Both it and the report written the previous year under the direction of the AEC, "The Nation's Energy Future," were very controversial. In the latter, the potential contribution of solar energy was declared by AEC Chairman Dixy Lee Ray to be "like a flea on the back of the nuclear elephant." What else could be expected from an effort chaired by the AEC? However, the Project Independence report acknowledged that "solar and geothermal power, while technologically feasible and increasing during the next ten years, would not have a major impact until after 1985."

Despite this more encouraging projection for renewable energy, the earlier AEC report had poisoned the well in Congress and resulted in several hostile hearings. Renewable energy advocates were convinced that the Nixon administration, led by AEC, were conspiring to keep renewable energy from realizing its full potential. Eggers was called to testify about the numbers we had included in our Solar and Geothermal Task Force reports. We had projected that renewable energy could have a large impact if it received more funding to advance research, development and demonstration. In our reports we had painted a rosy future for renewables based on optimistic assumptions that almost certainly would not be possible. But that is how the game is played in Washington. Solar energy advocates would quote from our reports to bolster their positions. My programs received substantial funding increases, from $4 million in 1973 to $50 million by 1975.

Energy Research and Development Administration 1975 to 1977

In 1975 many of my programs, along with most of my staff, were transferred to a new agency established by President Ford, the Energy Research and Development Administration (ERDA). My new ERDA title was Deputy Assistant Administrator for Solar, Geothermal and

Advanced Energy Systems; I reported to Assistant Administrator John Teem, a presidential appointee. The last part of my title included oversight of high energy physics and fusion research. Teem's boss was ERDA Administrator Robert Seamans, who had been the number two manager at NASA when I first reported. I have always been amazed that events and relationships kept repeating as my life went on. In almost all cases they resulted in important, favorable consequences. All my NSF programs transferred to ERDA expanded dramatically under Seamans' leadership.

Near the end of ERDA's first year, despite making progress in uniting the disparate energy programs he inherited from AEC, NSF and other agencies, Teem concluded that he could not satisfy all the very vocal protagonists telling him how to manage his part of ERDA. He found himself more and more at odds with the direction and budgets he was receiving from OMB. We were also in the midst of selecting a location for the Solar Energy Research Institute (SERI) mandated by Congress, a process that had become a political football. With these and other problems facing him, Teem resigned and was replaced by Bob Hirsch. I continued as Bob's deputy.

Bob Seamans was a great boss. I admired and tried to copy his management style. My first NASA meeting with Seamans occurred soon after I joined NASA. The subject was lunar exploration. I was a GS-13 invited to listen as senior management discussed how science experiments would be accommodated during Apollo missions. At one point I interrupted the speaker with a question. Seamans turned in his chair and, visibly annoyed, told me, "This is my meeting!" I didn't ask any more questions. Now, eleven years later, as a senior member of his ERDA staff I was invited to his Friday afternoon informal get-togethers when we sat around his conference table and he served us a glass of port. I'm not a big fan of port wine, but the meeting was a chance for all of us to be together, relaxed and able to discuss whatever issues we wished.

As a former Secretary of the Air Force Seamans could request an airplane, usually a King or Queen Air, and I would accompany him to out of town meetings if my presence was required. It surely beat a commercial flight, and refreshments were served by an Air Force enlisted man. Seamans entertained VIPs in his office with his own china. On one occasion, when we were wooing a Saudi prince by offering to help him establish a solar energy research institute similar to ours, I was assigned to be the prince's official host. With his entourage he came to my office at 20 Massachusetts Avenue where my secretary, Sophie Emami, greeted him in fluent Farsi. He was astonished and very pleased that we were so prepared. Sophie had married an expatriate Iranian engineer and had picked up the language. We used Seamans' fancy china to serve the prince fruit juice and cookies. We eventually signed a formal agreement to help the Saudis establish their solar research institute.

Progress was made in all applications of solar technology. Many demonstration projects, large and small, were built with special attention given to having one in every Congressional district so that the members could have their pictures taken cutting the ribbon. Larger projects were reserved to be dedicated by senators. For most projects I represented the government and explained the importance of the research. My files sent to the National Energy Research Laboratory in 1995 include a folder of press releases and my speeches associated with each project as well as records of all the research we conducted during the '70s.

Successful advances were made in the design and operation of solar heating systems for large and small buildings and wind turbines. Advances in other applications of solar energy, photovoltaic and high temperature solar power systems, were also moving on a fast track. Costs of building these systems were coming down, the primary objective of our research.

Operating the First Modern Wind Turbine

While at NSF I transferred funds in 1974 to the NASA Lewis Research Center in Cleveland (now named the Glenn Research Center) to design and build a modern wind turbine. Why NASA? Because a wind turbine blade is an airfoil, just like an airplane wing. I suspect that most people think a large wind turbine turns because the wind pushes it like a farmer's water pump windmill or whirligig. That's not so; as the wind passes over a turbine blade it creates lift, pulling the blade around just as an airplane wing provides lift in a vertical direction. When NASA was established in 1958, laboratories at the National Advisory Committee for Aeronautics were incorporated into the new agency. Important advances in airplane wing design had been made at Lewis while a NACA facility. Now, as part of NASA this research continued. Before the first wind turbine made under the NSF grant could be demonstrated, the program was transferred to ERDA.

Nonetheless, the day arrived to show the world how a future wind turbine, built at the Lewis Plum Brook Station, would look and operate. We hoped that on the chosen day there would be enough wind to turn the blades. Bob Seamans agreed to officiate at a ceremony to which the press and other dignitaries were invited. He would push the button to start it spinning. After a short speech came the moment; he pushed the button and the blade started to rotate slowly. A low thump was heard that grew louder with each revolution as the blades rotated faster. Thump,

thump, thump; we quickly pushed the stop button. With some embarrassment the ceremony ended. However, faulty design did not cause the thumping. After removing both hollow blades at the hub we found a flashlight left by a workman. At each revolution, following the laws of gravity and centrifugal force, it fell the length of the 50-foot blade, thump! Problem solved.

NASA Lewis went on to build many other wind turbines; one on the island of Oahu delivered over three megawatts of electric power. Today, based on that work, wind turbines are the best success story of the solar energy research we started in 1973, making a substantial contribution to the nation's electric power grid and growing in importance. Foreign manufacturers also benefited from NASA's work, as all its design and operating experience were available to anyone wishing to adapt them.

Soon after President Carter was elected Seamans talked to Carter's Chief of Staff, Hamilton Jordan, and volunteered to stay on until Carter could decide how he would proceed with his energy agenda. Seamans' offer was declined, and ERDA continued to operate for only a few months more. In total, it existed for only two years.

However, with the change of administrations Hirsch resigned. That left me, not a presidential appointee, in charge of one of ERDA's still functioning divisions. Responding to one of the high priority actions of the newly elected president, I had to quickly assemble a team to demonstrate his commitment to renewable energy and build a solar system to heat Carter's reviewing stand in front of the White House during the inaugural parade. We jury-rigged a system that didn't work very well, but the press never caught on and it was hailed by solar advocates, including those in the media. In addition, we built a solar hot water system and installed it on the roof of the White House West Wing. It never operated as well as advertised and was removed at the beginning of the Reagan administration. Neither Reagan nor any of his senior staff were ever solar energy advocates and research budgets declined in the years that followed his election.

1870s Photographs

Now a little change of pace. Before we were married Ann learned through a friend of an old warehouse in downtown Houston that formerly rented theatrical costumes and props and was about to be demolished. He had discovered under the floor boards a large collection of photographic glass plates. Together with her first husband they retrieved over 3,000 plates (3 x 4-inches) and stored them in their garage. When we moved her furniture to Maryland we brought the boxes along. Examining the contents in Maryland, we found that the photographs, ten to a bundle, mostly of individuals, had been taken in the 1870s, all carefully wrapped in yellowed newspapers from that time.

In 1977 I contacted the Smithsonian Institution Division of Photographic History and spoke to Eugene Ostroff, the division curator. He was very interested in the collection and agreed to accept the plates. As a condition of donating them I asked that we receive prints of each plate since I had heard that donations received by the Smithsonian were often stored away and never used. It wasn't called the "nation's attic" without reason. Because of our doubts about how they might be used we donated only half the collection. Every six months or so we received a box of 50 prints that we stored away. This story continues and ends in Chapter VIII.

Department of Energy - 1977 to 1978

Soon after his inauguration Carter dissolved ERDA and, following the direction in which President Ford was moving before he was defeated, established the Department of Energy (DOE). DOE consolidated all government agencies involved in energy research and regulation. Newly appointed Energy Secretary James Schlesinger now sat at the president's cabinet table. With energy concerns and problems elevated to this level, DOE became a huge bureaucracy with many new presidential appointees managing its various activities. Carter nominated Omi Walden for one of the assistant secretary positions, the management level below Secretary Schlesinger and his two top deputies. When confirmed she would have oversight of many of the programs I was managing at ERDA. But the Senate refused to confirm her. Her supposed sin: she was accused of showing favoritism to Governor Carter's brother Billy when gasoline allocations were made during the 1973 oil embargo. Billy owned a gas station and Walden, as director of Carter's energy department, was responsible for the allocations.

I am a staunch Republican, but as a career civil servant my political affiliation was never questioned. I was asked to stay on with the title Assistant Secretary for Conservation and Solar Applications (Acting). I ignored the "Acting" label and charged ahead to get all the programs inherited from other agencies, now consolidated in my office, functioning efficiently.

Coping as DOE Assistant Secretary

Management in a cabinet level agency is much different from managing in an independent agency such as NASA, NSF or ERDA. All decisions seem to have political overtones and repercussions, with one faction or another always ready to criticize. By chance, my new boss as DOE Under Secretary was Dale Myers, a presidential appointee, the former NASA Assistant Administrator for Manned Space Flight who succeeded George Mueller near the end of the Apollo program. I had briefed him many times during the last Apollo missions. In his new position we worked well together during the busy days starting DOE. Dale's immediate boss was Deputy Secretary Jack O'Leary, whom I had come to know in a former position at the Department of Interior. O'Leary reported to Schlesinger, another Washington bureaucrat with a long resume.

DOE began its existence coping with a new energy crisis. The bitterly cold winter of 1976-77 led once again to energy shortages and other problems. Intelligence estimates predicted a worldwide scarcity of oil and gas. Shortly after his inauguration Carter declared the energy problem to be the "moral equivalent of war" and vowed to take decisive action immediately. Before and after confirmation of his appointment Schlesinger spent almost every day testifying before one Congressional committee or another. Through 1977 and 1978 the nation's energy problems worsened as a result of the Iran-Iraq war. Prices at the gas pump peaked at over $3.50 a gallon. Carter, as had Nixon, focused on developing a National Energy Plan that would be accepted by all the involved parties to reduce dependence on foreign energy sources. Energy conservation and solar energy were to be the cornerstones of the plan, the responsibility of my new office.

While waiting to take up official DOE residence in the Forrestal Building, along with Dale Myers and other senior staff we had temporary offices in the Old Executive Office Building above Vice President Walter Mondale's office. It was in the White House compound and security was tight. I had to pass through a secret service check point to get to my office. My security clearances from my time at ERDA had been uprated to top secret and higher levels. I wore two badges that indicated these clearances and White House access by their color code and letters.

Until all my program offices were consolidated in the Forrestal Building, I continued to work out of the old ERDA headquarters at 20 Massachusetts Avenue, NW. I found it best to commute daily among the several offices where the staff from programs I was now managing still worked. This was made easier because I was entitled to a government limousine. The driver picked me up at home in the morning, drove me around the rest of the day, and took me home in the evening. For out of town trips, of which there were many, he took me to the airport and was there when I returned. The car had a cell phone, not very common in 1977, so I was in constant touch with all my offices. Pretty posh!

Sophie, my ERDA and now DOE secretary, rode along, taking notes and calling ahead for meetings as we dashed around Washington for the next eleven months. I had to meet and get acquainted with managers now reporting to me who were now one rung down in the bureaucratic pecking order from their previous positions. The meetings went well, and I believe we became a good team in a very short time. One such meeting was with Hazel Rollins, who had been an FEA appointee managing conservation research. She had a beautiful, very large office in the Old Post Office Building on Constitution Avenue, complete with high ceilings, fireplaces and wood-paneled walls. She had been expecting the turnover, so it went well and we worked together when she was given a position on Schlesinger's staff. Her office now became one of my many satellite offices.

For those without a background in Washington politics, Hazel was an attractive lady with friends in high places. Leaving the government during the Reagan and Bush administrations, she resurfaced and was appointed DOE Secretary during the Clinton administration. In between she married Jack O'Leary, which apparently added to her resume as she lacked any technical background. In the meantime, Jack died. Her short tenure as DOE Secretary was filled with controversy.

Life as a senior manager involved a great many meetings. Monday mornings started with Schlesinger's senior staff in his conference room in the Forrestal Building. We sat around a very long conference table with many chairs to accommodate the 30 or so who attended. Schlesinger, a rather dour fellow who seldom smiled, sat at one end smoking his pipe and carrying on quiet conversations with O'Leary, Myers and Al Alm, his senior policy advisor.

Occasionally a question would be asked of one us sitting around the table, but we generally just strained to hear what they were talking about. Even a seat close to Schlesinger's end did not mean one shared in the conversation. Based on my past experience, Schlesinger was not a typical senior manager. He wasn't interested in the day-to-day activities of DOE, which he left to O'Leary. He was better suited as a philosopher king than a manager of a new cabinet level department faced with multiple problems, not the least of which was just getting organized. At the few Congressional hearings I attended when he testified he seemed to enjoy jousting with committee members, providing long measured opinions in a deep voice, pipe in hand, when a short, straightforward answer would have sufficed.

After an unproductive hour or so Schlesinger's staff meeting was adjourned and we immediately reconvened in O'Leary's conference room. For the next hour the real business of DOE was discussed, action items assigned, the approach to be used for the many Congressional hearings discussed, and more. Because DOE now included all the federal government's energy regulatory agencies that in the past had operated independently, issues relating to their responsibilities consumed most of the time. Jack was a good manager, usually in a good mood considering the problems he tried to stay on top of. His style closely followed that of previous bosses. Because regulatory problems dealing with energy shortages took up much of the time, those of us with research and development programs did not receive a lot of attention.

Not to worry; as soon as O'Leary's staff meetings ended Myers began his. As Under Secretary he managed all DOE technical and research programs, a responsibility well suited to his background. He had come to DOE from his position as president of Jacobs Engineering, a large A & E company. Renewable energy, energy conservation, nuclear power, fusion and fossil fuel research were all part of his portfolio. In addition, he soon took on a completely new task, establishing the nation's Strategic Petroleum Reserve that would be located along the Gulf Coast. Dale, knowing of my background in geology and the oil industry, asked me if I would take the job of getting it started. It involved finding and filling large below-ground storage areas. The most likely candidates were salt domes of which there were many along the Gulf coast. Then he changed his mind and said he needed me in Washington. The job was given to Tom Noel.

Myers' meetings were much smaller than the previous two and included only assistant secretaries like me who managed research programs and a few other staffers. In his new DOE position he performed in a different environment from what he knew at NASA, where he wasn't a presidential appointee. As Under Secretary he was subject to all the political pressures and infighting of Washington, most of which were not concerned with managing technical programs. Ignoring these outside distractions, and using his NASA and private sector experience as a guide, he instituted monthly program reviews of all the research programs to keep on top of the problems we struggled with.

At the end of Myers' staff meeting my morning of sitting around conference tables was complete. In turn, my first afternoon action was to call a staff meeting of about 20 managers who reported to me. I reviewed the highlights, if any, of my three earlier meetings and the specific concerns of DOE top management. Then I took quick status reports on all our programs and problems. No matter how hard I tried to keep my meetings short, they were at least one hour long. This was the Monday routine I endured for eleven months, an example of why so little is accomplished in Washington. But coffee suppliers thrived.

A new administration must immediately develop a budget for the fiscal year that begins nine months later on October 1. It must also modify, to the degree it can, the budget submitted by its predecessor for the current fiscal year. Together these actions set the tone for the administration's agenda and allow the world to understand and critique the direction in which it will take the nation. At ERDA I had helped put together the Ford administration's budget for the current year, so I was required to testify before Congressional committees with oversight of my old and new programs. From March to May I made over 20 appearances before House and Senate committees, defending not only the Ford budget but laying the groundwork for the Carter budget. It was a very busy time.

Leaving DOE

By the end of summer 1978 Omi Walden had repaired her reputation, and new hearings were held on her nomination. This time she was confirmed and at the beginning of August took over my office and all the perks. I moved next door. She indicated she wanted me to stay as her deputy, but during the first two weeks she never asked me for help and I knew I had to leave. Carter had run on an anti-Washington platform. I had worked inside the Beltway for 15 years at three agencies and qualified as one of those he ran against. Perhaps Omi was reflecting the president's prejudices or didn't want to appear dependent on my experience when she met with her staff, Myers, O'Leary and Schlesinger.

I began an informal search for a new job, talking to friends with whom I had worked, and learned that NASA might be interested in having me manage its energy programs. I met with Al Lovelace, NASA's Deputy Administrator, and he asked me to come back. I accepted, and in August 1978 once again had an office at 600 Independence Avenue, SW, this time on the sixth floor.

Another short footnote: before her DOE appointment, Omi's management experience was limited. Under Governor Carter her office consisted of just a few staffers. She did not have a technical background; as I remember, her degree was in journalism. Suddenly she was in charge of a large staff and a budget approaching $1 billion a year, most of which was to be used for research, development and demonstration of diverse technologies. My former colleagues at DOE reported she was having a difficult time. Then they told me that one day, while riding on an elevator with Schlesinger, they had an argument and she was dismissed. Whatever the reason, she left DOE after a very short time.

Returning to NASA - 1978 to 1983

Just before I joined NSF in 1973, NASA Administrator James Fletcher established an energy office at Headquarters under the leadership of former astronaut Jack Schmitt. At NSF I had many interactions with Jack's staff, some not too cordial, as we competed for the same funds appropriated by Congress. NSF won that battle, and NASA's energy programs continued but at a reduced level. Soon Jack decided to leave and ran successfully for a Senate seat from New Mexico. With his departure NASA's energy work languished and the office was downgraded in importance. Thus, in 1978, there was an opening to fill Jack's position and reenergize the program.

In addition, Lovelace was concerned about NASA's overall future now that Apollo had ended and the human space flight programs that followed, Skylab and Apollo-Soyuz, were of short duration. He was especially worried that the Marshall Space Flight Center might not survive now that NASA was no longer building large launch vehicles, its special expertise. He hoped that with my background I could increase NASA involvement in energy research and development. I believed I could.

Taking up residence again at NASA, I inherited a staff of 15 that included three secretaries and an administrative officer. The latter, Mechthild "Mitzi" Peterson, was an important member. She had been a senior secretary in the NASA administrator's office before her transfer and promotion to work on Jack's staff. With that background she was able to steer me and our office through an organization that had changed significantly since I was there just six years earlier. It was no longer a "can-do" agency; it had become a typical Washington bureaucracy.

The professional engineers on my staff had varied backgrounds, having been assigned from different NASA Headquarters offices. Now, they were wondering what a new boss would have them do as reductions in force were rumored. With morale at a low ebb, my first months were spent getting everyone assigned to new projects that soon came our way.

Managing Solar Energy Projects

Before her dismissal Walden decided to emphasize demonstration and commercialization of solar heating and cooling in homes, government and commercial buildings, with the aim of rapidly transferring the technology to the private sector. I convinced Fred Morse, a former member of my staff but now reporting to Walden, that NASA could help this effort. We would assign available manpower at Marshall with tested engineering and management skills to both conduct needed research as well as oversee the demonstrations.

We quickly put together a team at Marshall that assumed responsibility for implementing DOE's Commercial Solar Heating and Cooling Demonstration and the Solar Federal Buildings Programs. The NASA engineers and scientists recruited for these efforts were not solar experts but had worked on NASA programs that required the same basic elements of science and engineering. Several hundred Marshall employees were soon engaged in designing and installing solar systems around the country and a few overseas. Marshall management viewed this involvement with mixed feelings, but with only space shuttle work in their immediate future they were generally pleased to receive additional funds that kept some of their work force engaged on a program that was immensely popular with the public.

Over the next four years I convinced not only DOE but Housing and Urban Development (HUD), the Agency for International Development (AID) and Department of Commerce (DOC) to send funds to NASA to implement their programs. By 1982 over 1,000 NASA employees, including some at the Jet Propulsion Laboratory (JPL), were working on energy projects with more than $250 million in funds transferred from other agencies. The job Al Lovelace hired me to perform had borne fruit.

Managing energy projects with funds from other agencies was a new way of doing business for NASA, and it made some senior managers uncomfortable. Projects ranged from developing electric and hybrid vehicles, a joint Lewis/JPL effort, to sterling engine and ceramic turbines for automobiles and trucks, to high temperature solar systems and magnetohydrodynamic power systems. Wind turbine development really took off under the direction of Lewis; wind turbines of many different sizes were built and operated at locations such as New Mexico, Block Island, Wyoming, North Carolina, the Columbia River and, as mentioned, Hawaii. The wind turbine program was intended to provide design, operating and maintenance experience that industry could use. It was successful, and most large wind turbines operating in wind farms today have benefited from technology developed by NASA in the 1970s and 1980s.

Three Mile Island Accident - Space Solar Power

Two other studies carried out by my office are of special note. After the Three Mile Island nuclear accident I suggested to the Nuclear Regulatory Commission (NRC) that NASA send a team to help understand why it occurred. It was believed that among other reasons control room personnel misinterpreted what was happening and took the wrong actions. I had once visited the control room at a commercial nuclear reactor power plant during my days at ERDA, and I was impressed with the many displays, switches and controls needed to keep the plant running. It reminded me, in a way, of NASA's Mission Control. I thought that a few on my staff who had some background in displays and controls could review the Three Mile Island control room procedures and provide advice on how to improve them. NRC thought that was a good idea, and I sent a two-man team to the plant. Our report was forwarded to NRC, but I don't know if any of our suggestions were adopted.

The other study involved examining the case for developing Space Solar Power Systems (SSPS). In brief, SSPS was based on placing very large (several square kilometers) solar collector systems made of photovoltaic cells in geosynchronous orbit. The solar energy collected would be converted to microwave energy at the same frequency that powers home microwave ovens, then beamed to Earth. At the terrestrial receiving station it would be converted to either AC or DC power and distributed on the nation's power grid, a straightforward application of known technology and energy conversion.

The concept had several attractive features. In geosynchronous orbit the collector system would almost always be in sunlight and could function as a base load power source as opposed to most Earth-based solar systems that stop generating power when the sun sets. In 1978 Johnson Space Center Director Chris Kraft, against the wishes of his bosses in Washington, Bob Frosch and Al Lovelace, lobbied the Texas Congressional delegation to appropriate money for an SSPS study. If such a power system were approved, it would require hundreds of payloads to build SSPS satellites in orbit. Space shuttle development was nearing completion (the first flight would be in 1981). In order for the shuttle to be a cost-effective space transportation system many launches per year were required. Potentially, SSPS was a made-to-order utilization of the shuttle.

Shortly after I returned to NASA Congress appropriated $20 million to be split between NASA and DOE to study the concept. I assigned three members of my staff to address the technology problems, using both NASA centers and contractor teams. Fred Koomanoff, who had been on my staff at ERDA and DOE, led the DOE studies addressing societal, regulatory and environmental problems. The result of our joint studies, completed in 1980, indicated that the laws of physics would not be violated by placing such systems in orbit 22,300 miles above the Earth. But there were many technological show-stoppers plus many other concerns uncovered during the DOE studies. The technology that would allow such a huge program to proceed did not exist, and the funds required to solve the technology gap would be enormous.

To be on the safe side, as SSPS proponents were very vocal in their support to continue pursuing its development, I decided to ask the National Research Council (NRC) to review our work. I asked Bob Seamans, who was a member of the NRC, if he would be interested in chairing the review. He agreed, and after several months of receiving briefings from my office and DOE, his committee released a report that said SSPS was an interesting idea but not economically feasible. It recommended another review in ten years when some of the technology problems, such as developing very low-cost, highly reliable photovoltaic cells, might be resolved through advances required in other space programs.

Ten years later a review of SSPS was conducted by the American Institute of Aeronautics and Astronautics and the Office of Technology Assessment. They came to the same conclusion as the earlier NRC study; the same technological show-stoppers still existed. Some at NASA were still strong proponents. They down-played the problems and lobbied, unsuccessfully, for continued funding. More recently the Department of Defense has studied SSPS hoping that such systems could alleviate the problem of supplying power to remote military units where bringing in fuel and generators is difficult and costly. Might SSPS ever become a reality? Perhaps, but I don't believe it will for the many problems we identified in 1980 that will be very difficult to solve. It will probably never be cost-effective compared to alternative Earth-based systems such as improved nuclear power plants and other electric power systems still on the drawing boards.

Energy Projects for AID

Funds received from AID were used to design and build a number of very practical and unique applications of solar energy. AID suggested projects for different countries lacking infrastructure and access to cheap and reliable energy. The Lewis team turned the suggestions into fully operational systems in a short time. By 1982 Lewis had renewable energy projects underway in 26 countries plus the US, including a few on indian reservations. Perhaps the most interesting and difficult was a project installed in Burkina Faso in central Africa.

An AID study of the Burkina Faso economy found that the women spent a major portion of each day hand-grinding millet, the food staple, and carrying water. AID reasoned that reducing the time for these chores would free women for other important work, including taking better care of their children. AID proposed to test the idea by building a large, centrally located solar-powered grain mill, water pump and storage tank in the small village of Tangaye. In a few months a complete system was designed, built and installed by Lewis engineers.

The mill's photovoltaic panels charged batteries which, in turn, powered the mill and the pump. There was sufficient power left over to provide an electric light, the only one for many miles around. After the women ground their millet the mill became a social gathering place. Families came and set up camps. Soon all available firewood was chopped down to cook the freshly ground millet. The neighborhood near the mill became an overcrowded, unlivable area. Always be careful and mindful of the law of "unintended consequences."

Another problem eventually solved the above consequence of building the mill. When harvesting the millet the women gathered all the seeds, including those that fell on the ground. They brought their baskets to the mill where they were weighed and the contents dumped into the grinder. Intermixed with the millet was the occasional stone which caused the mill to grind to a halt. We would have to send a technician from Lewis to get it started again. After a few such trips we told AID it wasn't possible to continue providing this support, and eventually the mill was closed down. Families soon moved back to their original locations.

A more successful solar energy application for AID was designing and building solar-powered refrigerators for use in developing countries. They were small units with attached photovoltaic panels charging a battery pack. Refrigeration made it possible to have available various kinds of medications, such as snakebite serum and vaccines which had to be kept refrigerated. This application of solar energy has been adopted in many countries.

Because our energy programs were not a NASA high priority, to improve their standings at the monthly status reviews for top management I decided to establish a small advisory committee of well known people. I convinced former NASA Administrator Jim Fletcher, then at the University of Pittsburgh; Omi Walden, working as a consultant but still with good connections in the Carter Administration; and David Cole, a well known automotive consultant at the University of Michigan, to be members. With their backing I thought NASA management would have to pay more attention to what we were doing. I would reserve the NASA Gulfstream airplane and with my advisory committee and staff travel to Lewis and Marshall to hold reviews. JPL reviews required using commercial flights. The committee's recommendations and critique of our research were useful, and I felt free to refer to them when reviewing my programs at the monthly senior management meetings with Frosch and Lovelace.

New Job Offer

I held semiannual reviews of all the programs at the NASA centers, so I knew all the senior management and met with them frequently to be sure they continued to support the energy work. Programs at Lewis were expanding rapidly. In 1978, when I began working with Lewis, its new director was John McCarthy. After several meetings he asked if I would be interested in moving to Cleveland to be his deputy, a position that was open. This was an intriguing offer. I knew the staff very well and had the highest regard for their abilities. I talked to several Lewis managers working on energy programs and all encouraged me to take the position. However, I decided not to do so. Tommy was completing college and Bruce just beginning. If I moved to Cleveland I might lose all touch with them. The Lewis job would undoubtedly have altered the rest of my professional career. In 1982 McCarthy was dismissed, accused of a conflict of interest in some of his dealings. As his deputy I would have assumed the director position, and perhaps Bob Frosch would have given me the job full time. NASA center directors are powerful people in the world of research, and Lewis was one of NASA's research jewels with unique facilities and a top-notch staff.

Colombia Vacation

In 1980, having been away from Colombia for 17 years, I returned with Ann for a vacation. We flew first to Bogotá, where we had reservations at the Hotel Tequendama. Our taxi trip to the hotel turned into an adventure. Just before we landed the Gold Museum, famous for its collection of pre-Colombian artifacts, had been robbed. All traffic into and out of the city was stopped. After a long delay we finally arrived at the hotel, giving the driver a nice tip for losing potential fares as we endured the long wait to be interrogated and searched.

After a few days in Bogotá we flew to Cartagena and stayed in one of the small hotels that line the beach between downtown and the old Hotel El Caribe. This was all new development since I was last here in 1963, extending westward from the city to Boca Grande. We rented a rather decrepit car and drove to Monteria, hoping to see how Figueroa and some of my old workers were doing. I found our former agent Marmol, who was now working for the city. He quickly spread the word that I was in town.

My prime objective was to see Figueroa and invite him to the US for a visit. But he had died two years earlier, so I missed the opportunity to talk to him. That evening I hosted a small fiesta for four former employees. I was amazed at how much they and other crew members had prospered in the intervening years. I learned that our lead mule man, Manuél Montez, now owned his own finca (a farm or ranch) and furniture factory. Our Dodge Power Wagon driver, Rosembérg Paternina, who had kept us supplied in remote areas he could access by rough roads and trails, told me he owned a small fleet of trucks. Our mechanic, Pedro Valencia, who had serviced our generator and maintained our outboard motors, had his own window installation business. Our other mule man, Julio Caro, also had a small finca. Friendly Sierra was now a cook at a girls' school; where else would one expect to find a ladies' man? Most had taken advantage of their employment with Mobil, and those who had saved some of their earnings had working capital to move from a world of remote camp sites to one of business.

Energy Program Terminated

Back at NASA, good progress was being made on all the energy projects we had undertaken for other agencies. Then in 1982, with President Reagan's election, our efforts came to a halt. Reagan had been elected on a platform that included moving many programs managed by the federal government to the private sector. Energy research, development and demonstration were among the first targeted for termination. His administration undertook to dissolve DOE, although it didn't succeed.

Jim Beggs, the new NASA administrator, agreed with this position. Soon after he was confirmed he told me to start canceling the hundreds of energy-related contracts being managed at the NASA centers. I accompanied him in his limo to DOE to tell them that NASA would no longer manage the automotive research programs underway at Lewis and JPL. These programs, in my judgment, were among the most successful applications of NASA know-how to advance technology in one of the country's most important business sectors.

The decision to stop energy programs was not well received at the centers. Because NASA appropriations were not sufficient to keep everyone working on space and aeronautics programs, there was the fear that some would lose their jobs. Beggs' deputy, Hans Mark, visited Lewis and told management and staff to stop working on energy projects or face dismissal.

We notified the sponsoring agencies of this development and began canceling our contracts. The interesting fallout was that research being canceled was with the very industry the new administration believed should be happy to fund themselves and not need any more government support. If not, then the research wasn't really needed to improve the nation's energy sector. In most cases, after government funding ended, the research was reduced or stopped.

An example was the work we were doing on electric and hybrid cars. In 1980 Lewis and JPL were making good progress in developing these vehicles. Major problems associated with controllers, instantaneous switching from batteries to an internal combustion engine, regenerative braking and more were yielding to their research. Every year we invited the automobile industry to review our progress. US companies turned up their collective noses at our research, but Japanese companies regularly sent representatives and were eager to benefit from the research paid for by US taxpayers. The result was preordained. Japanese car makers eventually offered the most choices in electric or hybrid cars and US automakers are scrambling to catch up. They would be way ahead if they had taken advantage of our research funded by US taxpayers.

It didn't take a rocket scientist to figure out that my talents were no longer needed at NASA. Hans Mark asked me to stay and join his staff. I politely refused and put out feelers in private industry. In my energy work I had become acquainted with Dan McDonald, one of the founding partners of Braddock, Dunn and McDonald (BDM), an engineering services company. He offered me an interesting position as vice president of their Houston office which was serving the oil and gas industry. He hoped I could also attract work at the nearby NASA Johnson Space Center. It would not entail a conflict of interest, as for the last four years I had not been involved in any space-related activities managed at JSC. Although I wasn't fully aware of how BDM did business (it was one of the so called "beltway bandits"), I accepted. Tommy had begun his career as an Air Force pilot and Bruce was enrolled at the University of Maryland. Concerns about not being nearby if they needed advice or counseling, that influenced my decision not to take the position in Cleveland three years earlier, were eased. In January 1983 I left NASA for the second time and we prepared to move to Houston.

Tom Begins His Air Force Career

During his high school days I had taken Tommy to fly with two friends, Rhett Turnipseed (the former VOA announcer you met as Turner) and Lou Divone, who were on my staff at NSF and ERDA. Both owned small planes. We flew around the Washington area and over Chesapeake Bay, and Tommy was allowed to take the controls. He must have enjoyed it. He received his degree at the University of Maryland, taking some Air Force ROTC courses in his last year, and soon after graduation joined the Air Force. Although it was never discussed, I assume he didn't want to follow his dad's Navy career; he had heard enough of my stories and decided to generate some of his own.

He began his career in 1981 at Officer Training School, Lackland

AFB, San Antonio. It was a new experience in military and self-discipline. At one point he had so many demerits that he was confined to base until he had worked them off. Apparently getting up at reveille, making his bed and then being ready for morning inspection took some getting used to. Ann and I attended the graduation ceremonies, and while there he showed us the barracks and how he saved time by sleeping on top of his sheet and blanket, which had only to be smoothed to be ready for morning inspection. With his classmates he paraded smartly before the reviewing stand and received his commission on May 6th. From Lackland he reported to Vance Air Force Base, Oklahoma, and began pilot training. Upon completing this phase, his hard-earned wings were pinned on his chest.

Tommy trained to fly the F4E Phantom at George Air Force Base, California. While there he married his high school sweetheart, Mary Beth Bowman, in April 1983. The wedding was held in Rockville, and Ann and I attended. His next assignment was to an active duty squadron stationed at Seymour-Johnson AFB in North Carolina, where he continued flying several versions of the Phantom.

Bruce Chooses His Future Path

After graduating from Robert E. Perry High School Bruce enrolled at the University of Maryland. He lived in an off-campus apartment and majored in philosophy. After two years and mediocre grades he dropped out and returned to live with his mother. I encouraged him to continue studying, and he enrolled at Montgomery County Junior College. At the same time he pursued his favorite hobby, golf. He played frequently and became a low-handicap golfer. However, he didn't continue his education. He found a job in Arlington, Virginia, as a waiter and later as a bartender. Over the years he worked at many restaurants and today is a bartender in Georgetown. He seems comfortable with his life and continues to follow his passion, golf. While we still lived in Rockville Bruce brought girlfriends by, usually a new one each time, to visit and enjoy one of Ann's dinners. I was always impressed with his choice of women and liked them all. But his romances, some that lasted for a long time, never resulted in a permanent relationship. When business takes me to Washington we get together for dinner and a chance to catch up on his life. In turn, he has visited us in Florida and taken advantage of my membership to play on the nearby golf course.

Meteor Crater, Arizona

Gene Shoemaker lecturing to astronauts at Meteor Crater, 1965

Laguna de Guatavita, Colombia. (Photo courtesy of Rudolf, Bogotá, S.A.)

LM mock-up and support vehicles at Meteor Crater simulation, 1965

Geology simulation for post-Apollo missions at Flagstaff, 1966

Astronaut training - Dave Scott, Neil Armstrong, Roger Chaffee, USGS instructor Joel Watkins, 1964

Waiting for Apollo-11 launch with Tommy and Bruce on Winnebago, July 1969

Tommy and Bruce at my briefing to Congress showing first Apollo-11 rock released from quarantine, September 1969

Apollo-15 astronauts - Dave Scott, Al Worden and Jim Irwin with a surprise presentation of autographed picture at Hadley Rille, 1971

Jack Schmitt and Gene Cernan training for their mission on Hawaii, 1972

American flag that went to the Moon on Apollo-17 presented by Chris Kraft, October 1981

Our wedding picture, March 1973

Our home on Dowlais Drive

One of Ann's special dinners for Bruce, Mary Beth, and Mother, 1974

Signing bilateral energy agreement in Moscow, September 1974

Ann's family home in Dundalk, 1974

Ann's mother Gertrude Keane and Ann's Sealyham, Shaun, 1970

Cutting ribbon with Senator Mark Hatfield for dedication of wood waste to oil plant in Albany, Oregon, December 1976

First modern wind turbine dedicated at NASA Sandusky, OH, facility, October 1975

With Marge and Tom at mother's 75th birthday celebration in Chicago, 1977

Bruce's Robert E. Perry High School graduation picture, 1979

Tom's commissioning, 1981

NASA/AID solar powered mill and water pump, Tangaye, Burkina Faso. Photovoltaic system (3.6 kw) behind huts. Mill in building to right of water tank, 1980

Chapter VIII
Private Sector Employment - 1983 to 1996

BDM Vice President

During my first three weeks at BDM I worked at its headquarters in McLean, Virginia, adjusting to its corporate culture. I could see some problems ahead, but my higher salary and benefits helped to overcome initial concerns. BDM was dependent on receiving contracts from the federal government, in particular from DOD. This required us to stay on top of forthcoming procurements and be prepared to submit winning proposals, and indeed BDM had put in place an elaborate process to do that. Top management spent much of each working day reviewing the status of proposal preparation.

Proposal writing was BDM's life blood. As soon as one contract was won, the process of winning another began. BDM offices located around the country were expected to support any branch that might need a key person to be included in a bid, demonstrating to a potential client that BDM had the required expertise. Most of my first three weeks were spent reviewing proposals about to be submitted to understand how BDM wanted them done. One of my major responsibilities in Houston would be writing winning proposals.

What I soon found was that often individuals, with special expertise, might be included on several concurrent proposals. This seemed to me to be a problem. What happened if more than one proposal was successful? I was told that management would "work it out in some manner" with the client's agreement. I was troubled. This practice could complicate contract negotiations if the key people bid weren't available. Apparently BDM competitors did the same thing. With lots of work coming in BDM was continually expanding the staff and hiring additional expertise. Beyond question, BDM was very successful and growing rapidly in size and stature when I joined.

Putting my concerns aside, I drove to Houston in my new company car to house hunt and become acquainted with my staff. I would be replacing the current manager, who had come under disfavor in McLean. One of my duties would be to dismiss him once the problems he had created were resolved. I didn't look forward to this, but it was a business decision already made in McLean. The current manager was aware of the problem, although not of his limited future, so it wouldn't be a personal rebuke on my part. Business is business; the bottom line rules.

House hunting was successful; I found a very nice home to rent west of downtown Houston in a quiet neighborhood only five minutes from my new office. We had put our Maryland home up for sale but no buyers appeared, so we decided to rent it for the time being. A family with one child was ready to rent by the time Ann had the furniture shipped to Houston. Not selling turned out to be very fortuitous for us.

The BDM Houston office was only a year old when I arrived. My staff consisted of five professionals and one secretary. Three of the five had the title of vice president, including the fellow I was to dismiss. BDM had found that the title was advantageous when soliciting new work, a philosophy of which I had been unaware. Presumably potential clients were impressed by dealing with vice presidents rather then someone with a lesser title. But business was booming and BDM was able to attract and pay professionals very good salaries. You can't knock success.

However, for the first time in my professional career my timing was bad. Houston had just entered what would be a long recession brought on by failures in the savings and loan sector and a downturn in the price of oil. Some of the new neighbors I met playing tennis had been Gulf Oil employees, but Gulf had recently been bought by Chevron which downsized its inherited Gulf Houston office. They were unemployed and job hunting in a bad market. Houston became known as the "see-through city" because so many of the shiny new office buildings were empty.

BDM management thought the Houston operation could weather this downturn even though the primary customer base was the oil and gas industry. We had several good contracts, and they were expected to be profitable. However, the biggest one was in trouble, a major reason I had been hired. My first order of business was to get that contract fixed. New work in support of the space industry would be the solution to expanding our operation.

Ann found a job at a nearby doctor's office, and we renewed some old friendships. Also, the Gilmores were now located in Houston, and Ann and Phoebe Gilmore became good friends based in part on their interest in cooking. Phoebe was compiling a cookbook of recipes gathered while living in Mexico that she was modifying so they could be prepared in a microwave oven.

Problems in Houston

The major contract held by my new office was with the Venezuelan national oil company, Petroleos de Venezuela. We were to develop a numerical simulation code for enhanced oil recovery, but progress was behind schedule. The problem was that my predecessor had underbid the cost to win the contract. By staying within cost it wasn't possible to apply enough resources to develop the rather complicated code on schedule. Venezuela sent an engineer to work with us, but he didn't have enough experience to provide much help. I concluded that his real assignment was to monitor the contract and keep his superiors informed of progress or lack thereof.

By this time I had dismissed the contract originator and added another staffer with numerical modeling background. I had no background in developing models. I could help only by monitoring progress against contract deliverables and finding ways to increase needed staff. Despite these problems work continued, and we conducted a few test runs of the model. Finally we had a program that showed we were close to achieving contract objectives, and we went to Venezuela for a demonstration run. The model didn't work as well as the Venezuelans expected, and they threatened to cancel the contract.

Returning to Houston I arranged to meet with my former ERDA neighbor and car pooler Harry Johnson, who now headed the DOE laboratory in Bartlesville, Oklahoma. His lab was developing a similar model. We agreed to cooperate and, assuming success, share the results. I thought I had solved the problem, and we continued to develop the model.

Meanwhile I made some contacts at the Johnson Space Flight Center and prepared to bid on a procurement. For my first attempt I put together a proposal that would need expertise from the BDM office in Austin. Austin turned me down. I went to McLean and explained the situation but received no help. So much for the company promise of easily obtaining mutual assistance.

I did convince McDonald that future contracts with NASA were possible. Based on my recommendation, BDM joined the National Space Club to elevate its profile in the civil space sector. Membership included all the major and minor companies involved in space programs. I attended the annual black-tie dinner in March 1984, sitting at the head table, and was introduced as the representative of the new BDM member. Ann sat at the table BDM reserved for invited dignitaries. But I was in Houston and all the action seemed to be in Washington. My NASA Headquarters contacts were taken over by a McLean vice president who went on to win some contracts. To add to my concerns, my immediate boss at BDM decided to move one of our best performing contracts back to McLean and I was told to terminate that manager. There was no explanation. My small staff became smaller.

1870s Photographs Revisited

When we returned to Houston we brought the remaining boxes of glass plates back with the intent of giving them to the Harris County Heritage Society. They were interested in seeing the plates, and I brought the remaining glass plates over for them to examine. The photographs had been taken at Barr & Wright, a well known studio located in Houston in the 1870s. Unfortunately, when Ann collected the plates, each carefully numbered, she couldn't find a ledger that might identify the subjects. I suggested that the Society try to find the ledger and in the meantime print some of the images and post them with the question: Do you know who this is? Perhaps someone would recognize the person shown in the photograph.

Having one's picture taken in the 1870s probably was expensive for the average family, not something one would do except for special occasions. Thus, some of the plates might be of early, prominent Houstonians. If they were identified, the Society would be able to add to the city's historical record. The young lady I spoke with at the Society thought this was a good idea, but she soon moved to Dallas and as far as we could find out nothing was done with the collection.

Return to Maryland - Consulting

Meanwhile, my association with BDM wasn't working out as anticipated. I started a new contract with Arco, but it was clear the recession in the oil industry was not about to turn the corner. Few NASA contracts were being awarded at the Johnson Space Center. I recommended closing the office and resigned after 18 months. I turned in my company car, left the office keys, and said goodbye to the remaining staff. I learned later that BDM soon closed the office. We notified our renters in Rockville that we were moving back.

Not selling the house had been fortunate. It would have been difficult to find as nice a house at a reasonable price in the same neighborhood if we had had to buy a new home. Back in familiar territory I renewed contacts with former colleagues and slowly got back in stride. Ann went to work at the same doctor's office. I was able to pick up some consulting contracts, and our financial situation settled down. What follows are a few stories based on interesting work during the next twelve years.

Costa Rica - Nicaragua Project

One of my first efforts was to start a project with Doug Rekenthaler, another recently resigned BDM vice president. We attempted to become involved in the Nicaragua conflict that made the headlines almost every day. Doug, a former Air Force pilot, had flown one of the planes that photographed the Soviet missile sites in Cuba. He had a number of good government contacts, including Major Ollie North who served on President Reagan's national security team.

We met twice with North in his small cluttered office located in the White House compound and described our plan to fly drones along the Nicaragua-Costa Rica border to monitor activity of concern to both the US and Costa Rican governments. North thought it was a good idea and gave us some contacts in Costa Rica. In March 1985 I flew to Costa Rica to discuss our proposal. I called the contacts North had given me and waited and waited but never received a reply. I enjoyed a few days in San Jose, a quiet, orderly city compared to some Latin American capitals with which I was familiar. In retrospect, it was an effort that probably never had a good chance to be successful.

Shortly after we talked with North he was promoted to lieutenant colonel and the Iran-Contra affair was exposed. North, along with his attractive blonde secretary and lawyer, became instant TV celebrities. During the interminable Congressional hearings the latter famously complained that he was being ignored by North's inquisitors and was being treated like a "potted plant." In my judgement, although North's actions embarrassed President Reagan, his behind the scenes activities led directly to the overthrow of a communist antagonist 700 miles from the US border.

Back to Colombia

Fortunately my consulting work was much more successful, and I soon had contracts with a number of Fortune 500 companies. One, with Chevron, took me back to Colombia. I read in a trade journal about Chevron activity in Colombia and on the chance that I might be able to help made a cold-call to Chevron headquarters in California. I told the operator my background; she immediately put me through to the Chevron vice president overseeing the Colombia work, who turned out to be a former associate, Cy Jacobs. Cy had been a consultant during my time in Colombia and had moved on to a more important position. When he heard who was on the phone he immediately picked it up and, before I could say a word, blurted out, "We have been looking for you!"

Recall that copies of the reports I made while working for Mobil in Colombia had been given to the Colombian government. Cy remembered that I had worked in the area where Chevron now held the former Mobil concessions. He expected that by reviewing the reports of our field work submitted to the government it would make the current Chevron field parties more productive. He tried, without success, to find them at the Colombian office where they should have been archived. This is not surprising since, based on my experience, many Latin American governments are poorly organized. I told him I had kept copies of all the reports and would be happy to send them to him. I hoped my generosity would be rewarded in some way. I was correct. Cy copied the reports and offered me a consulting contract that included working with his field parties in Colombia.

In early February 1986 I packed my old boots, Brunton compass and field clothes and flew to Cartagena. Larry Dekker, a Dutch-Canadian geologist in charge of Chevron's field work in the Sinú basin, met me at the airport. He was a very genial fellow with a great sense of humor, and we hit it off immediately. We drove to Monteria over the same roads in the same poor condition that Ann and I had traveled six years earlier. Larry and his team of some 20 geologists and helpers had rented a spacious downtown home, turning it into an efficient office with laboratories for fossil study, photo developing, drafting tables and more. I was impressed. He also had a fleet of four-wheel-drive vehicles. Mules had become an outmoded means of transportation.

Chevron had begun an extensive re-examination of the oil potential of northwest Colombia. Larry was planning to put five parties in the field to cover the whole area from Monteria west to the Gulf of Urabá and south to the Rio Verde. This was essentially the same area Jürgen and I had surveyed, but Larry would have many more resources at his disposal. He hoped to complete his work and make recommendations in two years. After a few days of traveling with Larry and other Chevron geologists, getting up to date on their work, I believed what he planned was possible. The rain forest was almost completely gone except for small patches on hilltops. Ranchers and farmers had bulldozed roads everywhere. I was told that only the area south of the Rio Verde, a little farther south than Jurgen and I surveyed, was still covered with rain forest and thus less accessible.

Over the next week I reviewed with Larry's team the mapping that had been done and traveled to some sites that Jürgen and I had never visited. It was now possible to drive almost anywhere we wished. One day, to examine a newly exposed outcrop in a road cut, we drove from Monteria to the San Juan River and back, a roundtrip distance of about 100 miles. This trip would have taken Jürgen and me over a week on our mules.

Since 80 years of previous field work and drilling had not found any major oil fields in this part of Colombia, why did Chevron believe it was worth a large investment to explore the area once again? It was a belief with which I fully concurred. After Mobil gave up its concessions, Colombia's flirtation with nationalizing the oil industry had vacillated back and forth. It was now in a good cycle that encouraged private investment. Oil might be found in the Sinú basin, since earlier efforts had never fully explored the area and as a result the geology and oil potential were still not completely understood.

In the early 1900s intriguing indications of oil deposits in the area had been found. In 1908 Standard Oil of Indiana drilled a well north of Monteria near the little river town of Lorica and struck oil. Apparently, based on sketchy records, the well encountered problems during drilling and was abandoned before it was completed. However, when additional wells were drilled near that location in the early 1900s, only minor oil shows were found.

When Jurgen and I visited the site more than 50 years later the original well, with a short section of pipe sticking above ground, was still oozing oil. A large patch of ground around the pipe was oil-soaked. Locals collected small amounts from time to time for various purposes. A few miles away, south of the oil terminal at Coveñas, another well had been drilled in the early 1900s that also had a good oil show. But, again, that well was lost because of mechanical problems. I took Larry to those sites so that he could see for himself that potential oil reserves might be found.

The most important discovery came in 1945 when Mobil's predecessor, Socony Vacuum Oil Company, discovered the small Floresanto oil field a short distance west of Monteria. It was an unusual discovery in many ways. Ten wells were drilled, and the field produced light, sweet crude for a short time. Production had ended many years before I arrived in Colombia. What made the discovery unusual was that it produced oil from a young (geological time) formation and the structure was very small and poorly defined. It was a geological anomaly. Regardless, oil had been produced in the Sinú basin, and through the years the knowledge of the Floresanto oil field spurred further exploration.

The two unsuccessful wells that Mobil drilled in the region in the late 1950s had been located on two structures recommended by Mobil geologist Bill Lawson. He had interpreted them as anticlines, usually a good place to drill for oil. Based on our field work Jürgen and I had reclassified Lawson's anticlines as fault structures formed when the region had undergone east to west squeezing in earlier geologic time. The dry wells, and our new interpretation of the geology that the many anticlines Lawson had mapped were fault structures, were the primary reasons Mobil decided to stop exploration in the Sinú. Without large, geologically attractive anticlines in the concessions the area became unattractive for further investment and, as noted earlier, the concessions were returned to the Colombian government.

Condensing this story, after one week with Larry's team I returned home. In April I returned to review progress and determine if the reports that Jürgen and I had written 30 years earlier had been helpful. On this visit I was joined by another consultant, Hollis Hedberg, a geologist of great renown who had worked for many years in Venezuela and Colombia for American oil companies. One of his areas of expertise was the identification of oil source rocks that he had studied in Venezuela and Colombia. This was of great interest to Larry, as to date exploration had not solved this question for northwest Colombia. There were tantalizing hints, but more field work was needed to identify formations that could be the source of oil.

This was my first opportunity to work with Hedberg although I was well aware of his background. A gentleman from the old school, he related with great ease to less distinguished colleagues. We stayed at the same hotel and had breakfast together every morning. I mention breakfast because Hedberg had a ritual. Each morning he would order "carne a caballo," a steak with one fried egg on top, sunny side up. At the end of the meal he asked the waiter for "papaya con dos semillas," a slice of papaya with two seeds. His theory, based on working for many years in the tropics, was that this breakfast, including the two tiny papaya seeds, had preserved his good health. I wasn't about to disagree; he was still very spry for an 80-year-old, climbing unassisted up and down outcrops during the week we were there.

Hedberg and I reviewed the latest results of the field parties. Good progress was being made, but Larry's team had encountered a troubling problem. He had been warned that narco-terrorists had moved some operations to northwest Colombia. Just two months earlier we had felt free to drive anywhere, but now we were told not to venture too far from Monteria, especially to the south and west. This wasn't a major problem, and we visited outcrops near Monteria that Hedberg hadn't seen for many years. I flew home after the week, planning to return in a few months.

A few weeks later I received a call from Larry. One of his field parties, two Colombian geologists working near the Rio Verde, had disappeared. There was no word of their being sighted or, as in most cases like this, a ransom request. He assumed they had stumbled on a terrorist camp in this remote area and had been killed. Chevron shut down its operation and gave up its concessions. Once more oil exploration came to a halt in the Sinú basin. If it holds sizable amounts of oil, they may never be discovered as Colombia continues to undergo the political turmoil and strife that has hindered development for most of its existence.

Vice President George H. W. Bush Election Campaign

In 1987 I volunteered to work on Vice President Bush's election campaign. I had been a precinct captain for Republican Congresswoman Connie Morella, who represented our Maryland district, but this was my first venture into national politics. In a meeting with Tony Lopez, a retired Air Force pilot who had the title of Deputy Director of Research for the Bush 88 Quayle Campaign, we decided that my talents were best suited to participating in Issue Groups that were being formed.

I became a member of three issue groups: Civil Space, Military Space and Latin America. Over the next year I attended many meetings of all three at campaign headquarters in downtown Washington. Our responsibility was to develop positions for the vice president based on our research. If his senior advisors agreed with our recommendations, they would become the basis for the campaign's responses when questions were raised as to how Bush would approach a specific issue when he became president. Each group was comprised of a dozen or so volunteers with varying backgrounds. In addition to preparing position papers I attended meetings with various lobbyists, some from other countries, who wanted to influence policies of a Bush administration.

As an example of what we did, recommendations were made on positions the vice president should take on civil space. Almost all were accepted. In a speech at the Marshall Space Flight Center in October 1987, for which we had supplied a draft, Bush unveiled a comprehensive list of space priorities to which he would be committed. He stated that he would support the construction of a replacement shuttle for Challenger, a somewhat contentious issue at the time. He endorsed the Mission to Planet Earth, focusing on environmental concerns proposed by an earlier NASA study. He would also support the development of a trans-atmospheric vehicle and construction of the Space Station. In his most ambitious statement he declared that "we should make a long-term commitment to manned and unmanned exploration of the solar system."

From our perspective, one of our most important recommendations was that Bush establish a National Space Council. This was not a new idea. President Eisenhower formed a National Aeronautics and Space Council in 1958. It was chaired by the president, but in 1961, in the Kennedy administration, Vice President Lyndon Johnson became chairman. The Council fell out of favor during the Nixon administration and was abolished in 1973. The Issue Group felt that Nixon and the presidents who followed did not have trusted advisors on civil space matters; thus there was a steady erosion of support for NASA programs. If civil space programs were to compete for administration support and resources, the president needed an informed advocacy council in his inner circle.

After Bush's election, and as a reward for my work on the issue groups, we were invited to one of the presidential inaugural balls. As I remember there were four at different locations. Early on the evening of January 19, 1989, with Ann's high heels in a paper bag, we boarded the Metro at the Rockville Twinbrook station and arrived just a few blocks from our destination, the Washington Convention Center. We were not the only ones who decided the Metro was the best way to get downtown. There were many others riding along; certainly the ladies dressed in long gowns were going to the balls. It was an interesting evening, listening to well known entertainers and a brief appearance by George and Barbara to say a few words and wave to the crowd.

Bush's handling of his civil space program agenda after winning the election was not a great success. He ignored our issue group recommendation that he chair the Space Council but chose instead to have it chaired by the vice president as had Kennedy. In theory having a Space Council close to the president was good; it functioned with some success during the Kennedy administration, laying the groundwork for his announcement of the Apollo program.

However, under the direction of Vice President Quayle and the small staff he assembled, the Council was a calamity and set Bush up for a painful defeat. The program developed by Quayle and announced by Bush on July 20, 1989, to return to the Moon and then go on to Mars, was ill conceived and poorly timed. With large budget deficits and NASA's major program, the Space Station, in disarray, it was easy for a Congress controlled by Democrats to deny the funding for what his Space Council conceived would be Bush's showcase program. A lot of time was wasted by many people attempting to convince Congress it was a great idea. The obvious conclusion: a National Space Council is only as good as its leadership and staff, and its existence does not guarantee that a president will receive good advice. Future presidential candidates take heed.

Boeing Space Station Proposal

As a consultant I often participated on a private company's "Red Team," established when bidding on a large government contract. The purpose of a red team is to critique the content of the proposal to be submitted and assess its likelihood of success. Some government requests for proposal run hundreds of pages in length and are very complicated. As a result, the submitted proposal might also be hundreds of pages or longer.

Making a company's bid attractive to a government agency's Request for Proposal (RFP) is a skill that is acquired after participating in many competitions and losing a few. Each agency has its own culture that colors how the proposal should be written. A company can spend millions of dollars writing a proposal to win a contract worth potentially billions of dollars, and a few hundred thousand more spent on a red team to assure success is usually money well spent. A company might establish "blue" or "gold" teams to review the proposal a final time after it had been modified to accommodate suggestions made by the red team. Almost all companies found this complicated process worthwhile, and I served on many teams.

I started red team consulting through the invitation of Mike Malkin. Mike had been the GE manager of the Air Force Manned Orbiting Laboratory (MOL) program until it was canceled in 1969. He was then hired by NASA as a senior manager on the shuttle program and was now retired. Mike and I worked together on many red teams and had a very good track record of helping companies submit winning proposals.

After working on a number of proposals I developed some skill in writing a proposal abstract. The abstract sets the tone for a proposal, highlighting the key elements; if well written and structured, it will make a proposal stand out from the competition. Senior government managers, who influence which proposal will be selected, may not have time to read the complete proposal, only the abstract. During my government days I had served on source selection committees and as a source selection official. I knew what should be included in proposal abstracts, what would catch the eye, what would set a proposal apart or, on the contrary, what would turn an evaluator off. The government request for proposal almost always limited the number of pages allowed in the abstract, so the abstract had to hit a home run in a very few pages.

Mike and I worked together on a wide variety of proposals for DOD communication, surveillance and weather satellites and commercial satellite applications. We also worked on a GE contract to study a space nuclear power system for NASA, and for DOE to clean up a nuclear waste site in Idaho and another to manage the nation's strategic petroleum reserve. The consulting fees we charged were very high. At times we put together the complete team of perhaps a dozen individuals; for others we were part of a team put together by someone else. The story below is the latter type.

At the end of June 1987 we were members of the Meridian Red Team Boeing had hired to review their Space Station Phases C/D proposal. Meridian was well known inside the Beltway as a contractor whose primary business was putting together red teams for their clients. Mike and I had worked with them before. The proposal Boeing was preparing was for the follow-on contract to the first Space Station contract it received in 1985, one of several that NASA had with various companies when developing its space station concept. The RFP had been released by the Johnson Space Center. For Phases C/D only one contractor would be selected. Until this time all space station contractor work was only paper studies. A win would put Boeing in position to manage major parts of the program when the hardware was built and placed in orbit and to receive billions of dollars in future years.

The proposal was so important that Boeing had contracts with two red teams, a rather unusual procedure. We arrived in Seattle a day after the other team that consisted mostly of retired NASA Johnson Space Center senior managers. Because or their backgrounds they were an obvious choice to review the proposal. Our team was given only a few hours to review the multi-hundred-page proposal so that Boeing management could receive both reviews at the same time. We quickly split the proposal into sections that could be reviewed by team members with special expertise on key proposal elements.

For the debriefing the room was filled with senior Boeing management and members of both red teams. The team of retired NASA managers gave their critique first and said the proposal was a winner with few changes required. Our team leader said the proposal was a loser and detailed all the problems we had discovered. Whom was Boeing to believe?

Boeing agreed with our assessment. It immediately changed the proposal manager, assigning a senior vice president to lead the rewrite, and sent a team to Huntsville to write a new proposal. From the Meridian red team I was the only one asked to join the Boeing team in Huntsville. We worked through the July 4 weekend and completely redid the proposal, finishing just hours before the submission deadline, and hand carried it to the NASA Marshall procurement office. The new proposal was the winner, and Boeing became the major space station contractor to work with NASA throughout the remainder of the program.

NASA Space Station Advisory Committee

With my background of studying how to place large payloads destined for space solar power satellites in geosynchronous Earth orbit, in the summer of 1982 I participated in some of the early Space Station planning at NASA Headquarters under the direction of John Hodge. John and I had worked together in the 1960s and early 1970s during the Apollo program when he was at the Manned Spacecraft Center and I was at NASA Headquarters. In 1982 John was at NASA Headquarters, given the mandate to lay the groundwork to develop a space station. But I left NASA in 1983, and my participation in the early planning was limited.

At the end of 1987 I learned that NASA was putting together a Space Station Advisory Committee (SSAC). I contacted Noel Hinners, NASA's chief scientist, another former colleague from Apollo days. He appointed me to the committee, on which I served until 1994. Committee members were selected from industry, academe, and former government employees. Many had worked previously at NASA. We usually held meetings every two months.

At our first meeting Andy Stofan, the newly appointed Space Station program director, asked us to provide advice and recommendations on needed technology, management and policy issues. He also wanted suggestions on how to provide stability to the program and to assure its utilization. During the program's first three years OMB and Congress had never fully funded the budgets requested by NASA. As for utilization, many in the science community thought research in micro-gravity wasn't a high priority. Instead, they lobbied NASA and Congress to fund other types of space research. There was an obvious need to improve interactions with those in the science community and convince them that a completed space station and its facilities would offer many unique advantages for researchers.

Those were the topics we focused on as we began our work, but we were soon asked to study specific technical problems. We took on the challenge, as we were happy to try to solve problems that the Program Office acknowledged needed help. Our Chairman Larry Adams, a retired Martin Marietta VP, split us up into seven small panels based on the background and interests of individual members. We studied such things as the threat of damage from orbital debris and inserting advanced technology seamlessly into the program in future years. As we were few, we usually served on more than one panel. Our responsibilities required frequent meetings followed by homework and report writing.

Meetings were held in Washington, at NASA centers involved in space station work, and occasionally at contractor facilities. At one point I accompanied Jack Kerrebrock, the chairman of one of the panels I was on, to testify at a Congressional hearing about a study we had just finished on extravehicular activities (EVA) required during space station construction and the use of robotics to complement the work of the astronauts. As the space station design matured, the amount of time that the astronauts would need to devote to EVA was growing rapidly, a troubling development seen by some members of Congress as an argument to cancel the program.

Our panel visited labs that were developing robots of various types and interviewed those involved. We concluded that robotic technology was not sufficiently advanced to help in the immediate future. Instead we recommended that NASA reexamine the design of the many elements that would make up the completed space station to reduce the amount of EVA time needed during assembly.

Being asked to testify at a Congressional hearing was an indication that the SSAC and its panels were considered important and relevant. Until it was disbanded in 1994 the committee continued to operate in this fashion, reporting to Stofan and his successors. You can read more about what the SSAC did in my book ISScapades: The Crippling of America's Space Program.

First Grandson Born

Updating Tom's Air Force career, from Seymour-Johnson AFB he was assigned in 1985 as an instructor at Shepard AFB near Wichita Falls, Texas. He had only four years of flying experience under his belt, but teaching from the back seat of the T-37 was a good assignment and from all the stories he told he was a very good instructor. I would guess he had just the right combination of compassion and discipline needed to turn a raw recruit into an accomplished pilot. After all, he had been in that position not so long ago.

While at Shepard Mary Beth gave birth in February 1988 to their first son, Travis. Ann and I flew to Houston and then drove to Wichita Falls. Travis, with his mother's dark good looks, was just one year old and crawling vigorously around their apartment. We also visited the base and saw the impressive flight line with hundreds of T-37s lined up neatly on the tarmac.

Pakistan

In the spring of 1988 I obtained an interesting contract. The US Agency for International Development (AID) had taken on the task of helping the Pakistani government evaluate its energy sector. AID was sponsoring a multifaceted Energy Planning and Development Project. The project, begun in 1983 with a budget of over $100 million, had as its primary objective increasing the country's domestic energy supply. To date AID had sponsored and completed several studies of Pakistan's energy sector, including a coal resource assessment and energy conservation analysis. Now it wanted an overall assessment of progress.

Our contacts at AID, including some I had worked with while developing the NASA/AID energy projects, offered Chuck Bankston and me a contract to perform the study. I put together a team consisting of five former colleagues with broad energy backgrounds. Our assignment was to review the results of the five years of energy studies already completed and to interview Pakistani government and private sector officials involved in energy work and other stake holders we deemed appropriate. We would provide AID with an analysis of progress made and recommendations for future work. Our draft report was to be submitted to AID and Pakistani government officials four weeks after we arrived in the country, a huge task to be done in such a short time, but I believed we could do it.

As none of my team, including me, had ever been to Pakistan, we had some research to do before stepping off the plane at the Islamabad airport. We were briefed at AID headquarters in Washington by AID staff, World Bank and AID contractors who had worked in Pakistan on the project. We were also given reports on AID's previous work and other background information.

Pakistan, a country the size of Texas and Louisiana combined, had a population estimated at 120 million in 1988. Its capital was Islamabad, but its major cities were Karachi and Lahore, located at opposite ends of the country. Climatically and geographically it was a country of vast differences, from subtropical flood plains where the Indus River emptied into the Arabian Sea in the southeast to the perpetually snow-covered mountains of the Hindu Kush in the north and along the western border with Afghanistan.

From an energy perspective Pakistan was definitely a developing country. Total electric power produced was only the equivalent of the power grid that supplies the greater Washington, D.C., area. And, as we would learn, pirate lines were everywhere draining off free power from the grid. Shortages and brownouts were common. Pakistan produced a small amount of oil and gas but was an overall oil importer. It had large coal deposits that were being mined. The problem was that 40 percent of the country's foreign exchange earnings went toward importing oil, a major burden on the economy. The untapped energy source was the sun which shone brightly over most of the country.

Compressing one month's work into a few pages, we traveled the length and breadth of Pakistan for three weeks. AID supplied the transportation, and on one occasion we traveled by train from Lahore to Karachi and back to Islamabad. Visiting the Northwest Frontier would have taken too much time, and southern Baluchistan was considered dangerous. Our visit to Quetta included in the original agenda was canceled.

We did spend two days in Peshawar, a crowded, somewhat dangerous city housing a million or more refugees from the war in Afghanistan. On our first night, while eating at a rooftop restaurant, we heard gunfire nearby. Jim Bever, our AID coordinator and division chief who directed our study and accompanied us on many of our visits, said, "Don't worry, it's probably some invitees celebrating a wedding."

We had no time to worry about safety, although there were many concerns. We were advised that, if involved in a traffic accident while riding in one of AID's vehicles, we should quickly leave the vehicle and stand unobtrusively at the side of the road until we were picked up. Pakistani crowds had been known to violently attack foreigners involved in auto accidents regardless of who caused the accident. It wasn't clear how we were to mingle unobtrusively in such a crowd, but we never had to try. Driving was hazardous; rules of the road were ignored. Stop signs and traffic signals were merely colorful decorations. Entrance into US government compounds was closely monitored. Before our AID vehicles were allowed to enter by crossing over thick, retractable concrete barriers, a security person would come out with a mirror on a pole to look under the vehicle for attached explosives. Then the barrier was lowered and we entered. We soon became used to this routine.

While in Karachi we took an interesting side trip. The beach west of the city was the center of an unusual industry, ship-breaking. Stranded on the broad beach were the rusty hulks of many large ships. Swarming over them was an army of half-naked men hammering and sawing away. A few had cutting torches, but most of the work was accomplished with raw muscle power. Piece by piece the ships were being dismantled. The disturbing aspect of our visit was seeing, from a distance, the living conditions of the workers. Higher on the beach ragged tents and old shipping containers served as shelters. What these men earned must have been little, but they were willing to live only slightly better than animals for a meager wage. At least they were employed in a country with high unemployment. Travel can be broadening if one ventures off the beaten path.

After three weeks of fact-finding we returned to the AID offices in Islamabad to put our report together. One team member, Art Rypinski, a former DOE employee and computer whiz, figured out how to convert AID's antiquated word processing system to one that would accept our laptop inputs. Among our recommendations was an outline for a conservation and renewable energy program to reduce the country's dependence on imported energy. Our report was not a draft as required in the contract, but much to their surprise a finished report; we were complimented on our recommendations.

On our final day Jim Bever took me for a quick visit to nearby Rawalpindi, a city teeming with street vendors and shops of all kinds. I wanted to pick up a few gifts for Ann and trinkets to remember my visit. I especially hoped to find a small prayer rug that would fit in my luggage. Jim took me to several rug dealers whom he knew to be reputable, but after visiting their shops I didn't find one I really liked.

However, at the first shop we visited I had noticed a small prayer rug hanging on the wall behind the owner's desk. It had an unusual square shape with white fringe on two ends and dark red and purple weavings in its pattern. We went back and I asked if it was for sale. The owner said it wasn't, so we turned to leave. "Wait!" he called." Do you have any pets in your home?" I said we had a dog. He said, "If you promise not to put the rug on your floor I will sell it to you." I promised it would be hung on a wall in our home, and that is where it is. When he took it down from the wall a bright area where the rug had hung remained on the wall. The rug must have been a favorite and hung there for a very long time. Back home, we had it appraised and it turned out to be quite valuable, woven in Iran many years ago.

One year after completing the above study AID contacted me again to perform a very different study. This time I was asked to evaluate the capability of a Pakistani institution, the Fuel Research Centre (FRC), for which AID through the years had provided $6 million to buy laboratory equipment. The FRC's major objective was to increase the use of Pakistan's coal resources. I agreed and asked a former colleague from my Apollo days, Bob Kovach, a professor at Stanford University, to join me.

In Washington Bob and I were brought up to date on FRC's programs. In addition, as I was somewhat familiar with DOE's Clean Coal Program, I asked for a briefing on its status in order to understand if the FRC could benefit from the technology. We were given a stack of reports and flew to Pakistan. Once there, AID added Fred Simon of the USGS to the team.

The FRC, located in Karachi, became an independent laboratory in 1984. With funding support from AID it purchased state-of-the-art laboratory equipment and received help in hiring a professional staff. Evaluating its programs was, to some degree, a follow-up to my first contract in 1988 that included reviewing Pakistan's coal resources and utilization.

Our approach to assessing the FRC's management and research capabilities was to first conduct interviews with the AID staff directly involved in FRC oversight. Then in Karachi we interviewed FRC staff and other Pakistani institutions with which the FRC interacted. We also examined the laboratory itself to determine how the equipment was being used and whether there was a need for additional equipment and professional staff.

We found that the FRC's performance was very good. Its leadership and management were performing their jobs well. The staff lacked some technical competence, but that could be fixed with a few key hires. We recommended that the FRC improve its interactions with the commercial sector that would be expected to use the technology it developed. We also made specific recommendations on future equipment the lab would need to perform all its research.

We were again complimented on our findings and recommendations. David Johnston, AID's manager of the overall effort to improve Pakistan's energy sector, asked me to stay and continue working at AID. I politely declined and Bob and I flew back to the States, mission accomplished.

End to the 1870s Photographs Story

Finally, in 1989, we received copies of all 1,838 glass plates we had donated to the Smithsonian. Many were photographs of babies held by proud parents and small children posed with toys of the era. However, before we received the final copies the Smithsonian notified us that they had selected photos from the Barr & Wright collection we had donated to be included in an exhibition of 32 early examples of children's pictures. *Beguiling the Babies: Photographing Children in the 19th Century* opened in July 1986 at the National Museum of American History, and Ann and I were invited to an early showing. My ploy worked. If I hadn't asked the Smithsonian to make the prints, I suspect the exhibition would never have occurred. Recently a Smithsonian historian of early photography, Shannon Perich, has written a book on the subject *The Changing Face of Portrait Photography: From Daguerreotype to Digital*. Unfortunately, during her research she was unable to locate the collection we gave to the Harris County Heritage Society. Thus her book will include only photos from our donation and other collections held by the Smithsonian.

Mexican Communication Satellite

The first red team contract Mike Malkin and I worked on was for a communication satellite to be built by RCA at their plant in West Windsor, near Princeton, New Jersey. Soon after that contract was completed, General Electric bought the RCA satellite division. Most of the RCA professional staff were kept on the payroll by GE, and Mike and I continued red team proposal reviews for them. In December 1990 I started an interesting contract to help GE sell a communication satellite to Mexico that would be designed and built at West Windsor. My Spanish and previous work in Latin America apparently were attractive to GE marketers.

Mexico had decided to have its own communication satellites in geosynchronous orbit that would service all of Mexico, Central America and parts of the US. Potentially it was a big market and in 1990 would make Mexico a major player in space and high technology. Prestige is an important facet of life in Latin America. At the end of 1990 Mexico released a request to build its first satellite, given the name Solidaridad. GE and Hughes, the two largest satellite builders in the US, were expected to respond. Mexico was a large market for GE products, including diesel locomotives, and the company was well positioned with a savvy sales force that understood how to influence Mexican decision makers. Influence is also an important facet of life in Latin America.

While the proposal was in preparation I was sent to Mexico to help GE's local staff lay the groundwork with the contacts that would influence the selection. I stayed at the Sheraton Maria Isabel just on the outskirts of the "Zona Rosada" in downtown Mexico City. Almost every day a GE contractor, Maria Arellano, would pick me up outside my hotel and drive me to the GE offices. There we plotted our strategy of meeting with government officials and institutions that would benefit from access to this new technology, such as the faculty and staff at the National University.

Our team put together the briefing material to be used. GE printed several attractive three-color brochures in both Spanish and English extolling its expertise and past associations with Mexico. In addition we placed full-page ads in all the newspapers. We chose as our logo a handshake, two arms with a bit of white shirt sleeve and coat showing, one hand a little darker than the other.

With the initial work completed I went home and returned in January to make the contacts and upgrade our sales pitch. It was an interesting and exciting time, as I had never before been involved in this kind of marketing. It was even more exciting after January 17, when I went back to the hotel at night and immediately turned on CNN to catch up on Desert Storm being fought thousands of miles away. Tommy was now stationed in Germany, so I was anxious to hear if his unit was participating. He didn't fly any missions during Desert Storm but later, during operation Northern Watch, patrolled the north no-fly zone established by UN mandate. He flew 10 missions from a base in Turkey. They were dangerous missions as he was usually tracked by Iraqi radar and shot at from time to time by Russian SAM missiles sold to Saddam Hussein.

After the proposals were submitted we continued to meet with government officials, attempting to follow the selection process, answering questions and handing out our brochures. Mexican procurement rules were not as stringent as those followed in the US. I also knew a member of the Mexican proposal evaluation team, Sam Fordyce, with whom I had worked at NASA during Apollo, so I was able to keep our GE team up to date on progress. The evaluation seemed to be on the up and up; we were confident of a win. Besides GE and Hughes a third team, Matra, had submitted a proposal, but it was not considered an important competitor.

Our confidence was misplaced; Hughes won the contract. Our team was debriefed after the selection, and we asked to see the Hughes proposal. Hughes had underbid us by $9 million ($143 million vs $152 million), but it was clear that Hughes had not included all the key elements that Mexico requested. If it had, its price would have been higher than ours. We pointed this out to the selection official, but the award had been made public and he wouldn't change. We later learned that Hughes had sweetened the pot to help influence the selection, a not unusual way to do business in Latin America.

But there was a happy ending to this story. Before I left Mexico the young lady who picked me up at my hotel, very attractive and from a prominent family, became engaged and I had lunch with the couple a few times. In November Ann and I flew to Mexico to attend their wedding. The reception was a colorful fiesta held in a large hall with hundreds of guests. We took advantage of the trip to do some sightseeing. Ann loves Mexican folk art, and we added some pieces to the collection we had purchased on our many previous visits to Mexico.

Second Grandson Born

Although Tom flew from an airfield in Turkey to monitor the no-fly zones in Iraq, his squadron's home base was Ramstein Air Base, Germany. Now he was flying the F-16 Falcon, the Air Force's hottest jet fighter. Near the end of this assignment Kyle was born in a German hospital. Ann and I were there for his birth in May 1991, and we entertained Travis at Kusel Castle on a hill overlooking the hospital while Tom was with Mary Beth. Kyle turned out to be a blonder Beattie than his brother Travis.

ENDOSAT Unmanned Aerial Vehicle

In 1991 Mike Malkin and I started a new company to develop a high altitude, long duration unmanned aerial vehicle (UAV). UAVs were being proposed by many aerospace companies to perform many different types of missions for the military and other users. However, the era of widespread application of UAVs was still in the future. Those that were flying didn't operate at high altitudes and couldn't stay airborne for long periods because of the limited amount of fuel they could carry. UAVs like the Predator drones used in Afghanistan had yet to reach their current level of maturity.

During Desert Storm operations, despite available satellites and other assets, commanders in the field had difficulty communicating with fast-moving armored divisions and coordinating air support for these units. We had the solution, a very high altitude UAV that could stay on station for months or longer, circling over a point on the ground. We named our UAV ENDOSAT for endo-atmospheric satellite, an aircraft that operated like a geostationary satellite within the atmosphere. Think of it as an eleven-mile-high antenna or a low altitude geosynchronous satellite that could act as a radio relay or carry sensors of various kinds to monitor what was happening on the ground.

Our UAV would fly at 60,000 feet or higher and didn't need to carry any fuel to stay on station. We would beam power to it using high frequency microwaves that would convert to DC power on the aircraft to run an electric motor and the propellor. At that altitude the line of sight footprint would cover an area approximately 500 miles in diameter. Just one of our UAVs would have allowed Desert Storm commanders to cover the entire battlefield. Sound like the better mouse trap? We thought so and put together a team to make it happen.

Microwave power beaming was not a new concept. During the technology assessments we made at NASA back in 1979 for solar-power satellites, using one of JPL's Goldstone facility large antennas we beamed microwave power to a receiver across the valley. In 1987 Professor James De Laurier, University of Toronto, designed and successfully flew a five-meter-wingspan aircraft powered by microwave energy that reached an altitude of 1,000 feet and flew for one hour.

Without delving deeply into the physics of this concept, here is a short primer. Several atmospheric "windows" permit the transmission of microwaves at specific frequencies with minimal power loss as they pass through the atmosphere. The above tests by NASA and De Laurier took advantage of one of these "windows," and microwaves were transmitted at 2.45 GHz. This frequency was chosen not only because of the window but because a power generating system was readily available at that frequency, the same one used in kitchen microwave ovens. However, if Professor De Laurier wanted to fly an aircraft larger than that used in his test and at higher altitude using that same transmitting frequency, he would need much more power. It would require a very large transmitting antenna and a large collector area on the aircraft to convert the microwave beams to usable power.

During one of my consulting contracts I learned that Arco Power Technologies, Inc. held patents for a microwave rectenna that would permit the conversion of microwaves transmitted at a much higher frequency (35 GHz) and higher power levels using another atmospheric "window." With these rectennas an aircraft could be built with a much smaller collecting area, and the antenna to transmit the needed power level would also be smaller. Thus, a larger aircraft and total system could be more easily designed if it used this technology. Other advantages could be realized such as having more power available for the aircraft's payload.

Mike and I started ENDOSAT, Inc. to develop this type of UAV. We established a team that included Arco Power to provide the power system and Aeronautical Design, Inc., led by Jim De Laurier, to design and build the aircraft. We later added Westinghouse Electronic Systems Group to provide sensors of various types as payloads.

De Laurier quickly built a larger test aircraft (twelve-meter wingspan), and it was successfully flown by remote control in November 1991. It was powered by a small internal combustion engine, but it was configured to carry microwave rectennas on a large disc-shaped extension of the fuselage. The flight demonstrated that De Laurier's designs could be successfully enlarged to the sizes we were beginning to advertise, 25-meter wingspans and larger.

Mike and I made 50 presentations during the next three years to market our UAV. We put together briefings specifically tailored to the needs of the US Army Southern Command, NSF's Antarctic Research Office, CIA, AT&T and many others, describing our UAV as the ideal platform for communication, environmental monitoring and surveillance. We received encouragement but always with questions about cost, technology readiness and others.

We had good answers for most of the questions; we had done our homework. But there were some we couldn't answer at the moment, such as having frequency allocations reserved by the Federal Communication Commission and the use of controlled airspace by unmanned platforms. We said we were addressing all concerns and that we were planning an all-up test with a microwave power system and De Laurier's twelve-meter-wingspan aircraft carrying the Arco Power rectennas. We made this prediction because the center director at NASA's Wallops Flight Facility at Wallops Island, Virginia, had agreed to allow us to conduct the test at his facility.

But the best laid plans sometimes run off the track. In October 1994 Mike had a heart attack and died. With the family's approval I arranged to have him buried at Arlington National Cemetery. As an 18-year-old radio operator in the 5th Marine Division Mike had landed in the third wave at Iwo Jima. He had many stories to tell but never dwelled on his service. On November 1, 1994, with full military honors, Myron S. "Mike" Malkin's ashes were interred in the Arlington Cemetery Columbarium. Without my partner it would be difficult to continue.

Then the wheels came completely off the track. In 1995 E-Systems Inc. bought Arco Power. Primarily a government and DOD contractor, E-Systems wasn't interested in continuing a relationship with ENDOSAT. Without the rectenna patents that they now held I couldn't continue and dissolved the company.

History of Solar Energy Research

Years earlier, in 1985, I had received a contract for $35,000 to write and edit a book on solar energy. It was to be the first in a ten-volume series recording all the research, development and demonstration of solar heat technology that had been done since 1973 when I began managing the program at NSF. However, my two co-authors, Chuck Bankston and Fred Morse, were too busy at the time to contribute, and the book languished. It wasn't of great concern, as I was also very busy. I had made a start on my portion, putting together an outline, some notes and source material. In fact, as a pack rat I had many boxes of reports, memos and other background collected during the ten years I had been directly involved in solar energy research.

Now, in 1995, my two co-authors and I had time to start writing in earnest. My boxes of reference material, six in total, had made the round trip to Houston and were in our basement in Maryland. I brought them to Chuck's office in D.C., where they would be available to all of us. I would work on and off at this office for the next year, writing my section and editing Chuck's and Fred's drafts. All of my part was written in longhand, and Chuck's secretary turned it into a finished manuscript.

Chuck was the overall editor of the ten volumes. The other nine, each addressing a different aspect of solar heating, had already been written by teams of researchers and submitted to MIT Press for publication. Although our volume was listed as first in the series, it was the last to be completed. As it turned out, this was an advantage. We used what had already been written as a guide to assure that our volume tied everything together. Attempting to review in one volume 20 years of research and development funded by over $7 billion of tax dollars was a daunting task. We believed we had met the challenge and at the end of 1995 submitted our draft to MIT Press.

Visiting Family and Friends

While waiting for MIT Press to respond, I will wrap up this chapter of my life with a few stories describing enjoyable times visiting family, friends and Ann's Irish relatives that took place during the years we lived in Maryland and Florida. Ann's younger brother, Edmund, was a teacher who lived in Northern Ireland. He was married but had no children. Her older brother, Gerald, was a constable and with his wife Patricia had three children, Gerald Jr., Richard, and the youngest, Adrienne. All three would come to the US to work during summer school vacations. The boys, both robust and over 6' 2", had a contact in Texas that employed them in construction projects. Adrienne usually worked as a waitress at restaurants in the Boston area or on Cape Cod.

Gerald and Pat visited us once in Maryland, and Gerald spent some of his time at local police stations. With his Irish gift for storytelling he undoubtedly regaled the officers with tales of policing in Ireland. For a time he was stationed on the border with Northern Ireland when Ulster was engaged in a civil war. I had heard some of his stories, comical and dangerous adventures while patrolling the border. His Maryland police audiences must have enjoyed his visits, as he returned to Ireland with badges and other law enforcement souvenirs.

Richard stayed with us once for a few days when returning to Ireland from Texas. His visit included a weekend when the Washington Redskins were playing at home. I held two season tickets to the games, usually taking Bruce or a friend, but this time I took Richard. He knew little about US football, so I gave him a crash course on the rules and star players he would see. It was my hope that when we arrived at the stadium he would be interviewed by Johnny Holiday, a local radio celebrity who broadcast from a trailer outside the main gate. As fans entered the stadium he asked for their opinions of how the game would turn out.

I told the attendant outside the trailer that I had a visitor from Ireland who was attending his first US football game. It worked. Richard was invited in and interviewed. Based on my briefing he came across as very knowledgeable; he even discussed the playing ability of Joe Theisman, the Redskin quarterback. At the end of the interview Holiday asked Richard the name of the uncle who had brought him to the game. Richard stammered, forgot my name, and made one up. The interview was taped and I obtained a copy and sent it to Richard in Ireland. When Ann and I visited again we made him play the tape for the assembled guests; all enjoyed listening to his embarrassing memory lapse, including Richard.

We went back to Ireland several times, including the 1987, 1989 and

1993 weddings of Gerald Jr., Richard and Adrienne. All were marvelous large affairs of family and friends, each highlighted by an Irish-style roast at the reception after the wedding. At the long head table the dignitaries, priests and members of the wedding party each tried to outdo the other with funny and sometimes naughty stories at the expense of the bride and groom. Richard's wedding was, perhaps, the most notable. The high mass was celebrated by five bishops and priests related to his bride. Two came from the US for the occasion.

Gerald Jr. and Richard became noted lawyers. Their mother's brother, Liam Hamilton, was the Chief Justice of Ireland's Supreme Court, and Richard eventually pleaded cases before it. Gerald Jr. made his name and fortune representing famous artists and performers, including U2. Adrienne Fitzgibbon, after a career in banking, is the proud mother of three boys, the last additions twins.

Ann and I traveled the length and breadth of this beautiful country, surviving many miles of narrow country lanes driving on the wrong side of the road. On my first visit to Ireland as a young naval aviator in 1954, with a different Irish lass sitting beside me, I could not have imagined that I would eventually have these stories to tell.

We also made several trips to Europe to visit with the Haffers and Tommy when he was stationed in Germany. We sometimes combined those visits with other sightseeing, including a very enjoyable trip to Portugal in 1998, 50 years after I was there as a midshipman. We spent one week with the Haffers at their favorite pensióne on the shore of Lake Como, visiting many interesting sites, sailing the lake and enjoying the food of northern Italy before returning with them to their home in Essen. During our visit to Lisbon and Germany they came to the US and stayed at our home housesitting our dog Biscuit, Pretty Kitty and cockatiel Coco.

The Haffers came frequently to the US to join cruises starting in Florida. Jürgen's fame as a naturalist made him a sought-after lecturer on cruises to the Amazon and Caribbean. We often spent two weeks in the fall at Cape Hatteras, renting a home just behind the dunes. One year Jürgen and Maria joined us to experience and enjoy some of the best beaches in the world. Jürgen's death in 2010 brought an end to our close friendship although we continue to exchange news with Maria and their children.

Chevron crew in Monteria, April 1986

With Larry Dekker after a hard day's work

Larry and Hollis Hedburg at 1908 oil seep

Area that was once Colombian rain forest during my time in 1958

With driver of typical highly decorated Pakistani truck, May 1988

1870s photo of small boy in fireman's hat included in Smithsonian exhibition

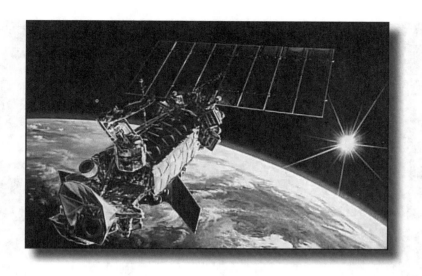

One of the DOD satellites (DMSP) I worked on with Mike Malkin

ENDOSAT aircraft on runway for first flight, April 1992

Tom and Mary Beth at Mom's Florida home, 1984

Bruce with Mom and Ann at our Maryland home

At the George H. W. Bush Inauguration Ball, January 1987

Ann's brother Gerald and wife Pat

Tom and Mary Beth in F-16 cockpit at Ramstein Air Force Base

Ann and Biscuit enjoying the breeze at Cape Hatteras

With the Haffers at our favorite Ocracoke restaurant, the Back Porch

With Harvey and Bobby at Isabel's home 1995

Vacation in Portugal

Vacation in Luxembourg

With the Haffers at Lugano on Lake Como

With Tom and Marge at mother's 95th birthday in Brookings

Marge and Turk at the 95th

Dowlais Drive in winter - why we moved to Florida

Chapter IX
On the Last Lap - 1996 to ???

When I closed the ENDOSAT project and it became more and more difficult to find interesting consulting work, I began a new career as an author. Also, two of the doctors for whom Ann worked were retiring at this time. Add to these considerations a particularly cold, snowy winter; shoveling snow and scraping ice from our windshields was less and less appealing. We decided we had sampled enough of Washington life inside and outside the beltway and began searching for a warmer place to relocate. I began taking short trips to the Eastern Shore of Maryland and as far south as Florida looking for places where we would enjoy living.

Finding a New Home

In 1995 I finally found a location just south of Jacksonville, Florida, where I believed we could live comfortably. To confirm my choice, with Biscuit in the back seat Ann drove down with me to see if she agreed. I had been searching the area with a very pleasant real estate agent who knew all the builders, their pluses and minuses as well as the new neighborhoods. She had shown me many existing homes up for sale and show houses at the new developments. A building boom was underway in northeast Florida. A development in northwest St. Johns County, Julington Creek Plantation, was particularly attractive. Ann agreed.

As morning people we selected a site on which the breakfast room windows would face east with a view of a lake that came to the edge of our property. We chose a modest four-bedroom (one became my office), two-bath design with a two-car garage. The latter would be real luxury, as we had parked our cars outside for 23 years in Rockville. Once the house was finished we planned to build a pool on the east side, screen it in and install solar panels on the roof to warm the water. On the day we signed the contract the builder offered for the first time steel framing at no additional cost. We accepted; there would be no worries about dry rot and termites!

Compared to some we knew who had experienced problems when building their new homes, ours went up without a hitch. I drove down twice to check on progress. I wanted to make sure that the steel framing was well grounded; I didn't want to be drilling and hanging pictures on a wall and get a shock, literally and figuratively. Both the realtor and the builder sent us pictures every week, from the uncleared lot to the finished house. The pool became one of the most used features. We swim in water heated by the sun from late March to early November. We could use the pool year round if our old bones would tolerate water temperatures in the 70s.

We put our Maryland home up for sale, and this time it sold on almost the first day. It was fairly priced and the house was in excellent condition; among other amenities it had a completely new up-to-date kitchen. We packed our furniture and 23 years' accumulation of who knows what and moved to Florida the first week of July, 1996. Before moving I had sent all the boxes of records related to my energy work to the National Renewable Energy Laboratory library in Golden, Colorado (formerly the Solar Energy Research Institute I had helped establish), packing and picking them up for the last time. My back thanked me. I hope the contents will be useful to future historians who may wonder how we spent over $7 billion between 1974 and 1995. However, the many boxes of NASA records containing sources and references for my next two books made the trip south again.

First Book

After we had moved into our Florida home, MIT Press sent me the proof copies of our solar book for review. One of the most difficult decisions was selecting the photographs and charts that would best help a reader understand the management and challenges of the solar heat program. Some of the problems have already been noted in Chapter VII. I had kept a large collection of this material, but the high cost of printing limited what could be included. When cost is a factor, compromises are reached. Although I would have liked to include more, MIT Press was generous and the book is well illustrated.

After release in 1997 several reviews were printed, a few not too flattering. One reviewer wrote that most of the book was "long-winded and laborious." Another warned readers of the "five-page glossary of acronyms." Addressing this comment, as I was forced to do for my next books, long lists of acronyms are required if one wishes to write authoritatively about large government programs. There usually are so many individual parts with long names that to spell them out each time would turn readers off. The acronym lists will also help when comparing my histories with other accounts which might identify a project only by its acronym.

Some reviews were kinder. "The last two chapters are more rewarding for students of technology." In these chapters we condensed the results of the research and provided lessons learned. Another reviewer said the book "provides the most comprehensive record of the U.S. government's solar thermal program" and also liked the plan we outlined in the last chapter, a national energy program that would include renewable energy.

I was quite satisfied with my first attempt to write and publish a book. I don't know how many complete ten-volume sets or individual volumes were sold. Fortunately I didn't have to worry about MIT Press turning a profit on the series. My only concern was to publish a truthful account of a program I had managed in the hope that the lessons we learned would be useful to others who might follow in our footsteps. However, the experience encouraged me to begin another book that I had been planning for 30 years. But, as a new resident of St. Johns County with an engineering background, I was enlisted by neighbors and new friends to become involved in local issues and politics. I was soon embroiled in many contentious issues while at the same time writing two more books.

Getting Organized and Local Politics

Soon after we moved in a neighbor asked if I would help to have the main north-south road, State Road 13 (S.R. 13), designated a Florida Scenic Highway. The road, running along the east shore of the wide St. Johns River, was overhung with centuries-old oak trees and passed through quiet farmland and horse ranches. A group of local residents were organizing a committee to petition that it receive the designation which they felt was needed to save the character of an area in danger of being overrun with strip malls and high-density housing developments. In 1988 Governor Bob Martinez designated the giant oak trees growing along the road since the Revolutionary War a "Florida heritage," and they are now nicknamed "the governor's oaks." A formal designation as a scenic highway was the logical next step. I agreed it was a worthwhile goal and joined some dozen neighbors to start the effort.

For the next ten years we labored, scheduling monthly meetings to obtain support for the designation from the county commissioners and the Florida Department of Transportation. We finally succeeded. The county administrator appointed a staffer to help us complete all the paperwork, and a state employee was also assigned to help us find our way through the bureaucratic tangle. As I write this the section of S.R.13 beginning at the northern border of St. Johns County with Duval County and running south for some 20 miles is now designated the William Bartram Scenic and Historic Highway, and a management plan has been developed. The roadway is named for the famous 18th century naturalist who studied the area more than 200 years ago. Come visit and enjoy!

When the St. Johns River was designated an American Heritage River in 1997, along with some eight or nine other rivers scattered around the country, I volunteered to help. Its designation was not supported by some in the belief that the river would now be more closely regulated, adding onerous requirements. That was not the case; instead it began to receive additional resources to help maintain and restore the river to its earlier health and beauty. Over the next three years, as the St. Johns County representative, I served on several committees that were formed to address different aspects of the initiative. I no longer am involved, but the work of the initiative continues with participation of all the jurisdictions that border the river.

As Thomas "Tip" O'Neill, former House Speaker from Massachusetts, was quoted as saying, and he should know, "All politics are local." Small-town politics can be very messy and at times nasty. Northeast Florida was in the midst of a residential building boom when we moved there. Our particular development, a few hundred homes at the time, was designed and approved to have some 6,400 detached homes and townhouse units; it would be a small city in itself. We were aware of this but weren't concerned. Our part of the development was on the edge of the main section and had opened up just the year before. There were only four other homes on our cul-de-sac, and the lots were generous. We didn't think we would be crowded, and we weren't. Over 700 homes were eventually built in our section, The Parkes. But there were some real problems associated with other proposed new developments in our area.

St. Johns County is managed by a board of five elected commissioners. I soon learned that the board approved almost all the new development projects it reviewed, usually by a vote of 4-1. The one negative vote was the commissioner who represented our district. I should mention that our development had been approved more than ten years before we arrived. It was the first large-scale project proposed in the county, and its impact was just beginning to be felt in 1996. It had changed ownership several times, and buildout had been very slow because of the developers' financial problems.

Our new friends and neighbors were very concerned about what was happening and enlisted my support. I met the commissioner for our district and offered to help him analyze the impacts of what the commissioners were approving. Anyone who took the trouble to review the applications for new development could see that the county was headed for trouble. Developers were not required to provide the infrastructure needed to support their projects, so roads, fire stations, schools and other needs weren't being funded or built; the impacts were being passed on to existing county residents.

Over the next few years the board approved over 40,000 new homes in our area alone, always with the 4-1 vote. Board meetings were held every week. During public comment periods I often provided detailed analyses of projected traffic congestion and school crowding that would be created by the new developments, all to no avail. For example, the potential 40,000 new home owners would have to commute using one two-lane bridge (later widened to four lanes but still congested) over the waterway that separated St. Johns County from Jacksonville, where most of the jobs were located. The new high school designed for 1,500 students was built three years after we arrived; the day it opened over 2,000 students registered. Schools at all levels were forced to place dozens of temporary classrooms on their campuses to accommodate the unplanned-for students.

Although the county is divided into five districts, each commissioner is elected by the vote of all county residents. When we first moved in a typical candidate for commissioner might have a campaign chest of $10,000. Soon some candidates were raising over $100,000 to advertise their qualifications. The majority of those funds were provided by developers and related businesses; some large contributions came from outside the county. Those receiving the large donations were of course committed to approving new development but had the resources to blanket the county with colorful advertising touting their good intentions. The problem and its solution were clear; there had to be a "take-back-the-county" movement.

We decided we had to reverse the 4-1 vote. I declined to run for the position of commissioner but worked with those who did run on platforms to better manage the county. One by one, as commissioners came up for re-election every four years, we brought in new faces; the good-old-boys network was losing power. As new commissioners were voted in, the board would vote 3-2 for development, then 3-2 to manage development in a sensible manner. A 4-1 vote to manage growth and county budgets carefully has not yet been achieved, but it is close.

By the time we were able to elect a board to conservatively manage the county finances, previous commissioners had exhausted the county's borrowing authority by failing to make developers meet their obligations, passing on the shortfall to taxpayers. It became clear how the majority of voters wanted county matters managed. Commissioners who originally sided with developers became more selective in how they cast their votes, occasionally voting with the new majority so they could claim to be listening to the voters, but one by one they were voted out.

I was appointed to several county committees. One sought to make it more business-friendly, a sorely needed evolution as it was essentially a bedroom community for businesses in Jacksonville. Little more than ten percent of the tax base came from businesses, an unsustainable situation if the required infrastructure were to be put in place for all the approved homes. We couldn't unapprove all the housing developments previous commissioners had approved, so we just chipped away adding needed infrastructure in the years that followed. The recent recession and downturn in home building have given us a little breathing space. Additional schools were built, including two high schools, to reduce overcrowding.

I spent uncounted hours working as a volunteer on many issues, so many that the local newspaper, the *St. Augustine Record*, asked me to write a column on the rewards of volunteering. I have reduced my involvement in local politics. But the old special interests are still lurking and trying to make a comeback. Perhaps we have turned the corner, but I don't believe development interests have changed their stripes. Building homes in St. Johns County is very attractive for many reasons. I have left it to others to fight the good fight. In the meantime I have concentrated on writing about subjects of which I have some expertise.

Tom's Participation in War with Yugoslavia

Since I last wrote about Tom and his family he finished two tours, one on the staff at Ramstein and another as an F-16 instructor at Luke AFB west of Phoenix, Arizona. In 1998 he joined an F-16 squadron at Spangdahlem AFB in western Germany near the border with Luxembourg. In early 1999 Yugoslavia invaded its neighbors and NATO responded. Tom's squadron was sent to Aviano, Italy, to participate in the action just across the Adriatic Sea.

During this deployment Tom flew 29 combat missions supporting NATO flights bombing targets in Yugoslavia. His most exciting flight, his second, occurred on the night of March 28. While he was returning from a support mission an Air Force F-117 Night Hawk stealth fighter was shot down. The pilot, Lt. Col. Dale Zelko, ejected safely but landed in enemy territory. He radioed his position and tried to stay one jump ahead of Yugoslavian forces seeking to capture him. A US Air Force Para-rescue Team was quickly assembled to bring him home. Tom and his wingman were alerted to the rescue attempt and stayed over the site to provide support if needed. For the next nine hours they orbited overhead, refueling seven times from a tanker. The daring rescue was successful, and for his part Tom was nominated for the Distinguished Flying Cross.

Second Book

Soon after my first book was published I began an outline for the book I really wanted to write on how we overcame many obstacles to provide science payloads of the highest merit and value for Apollo missions. There were other reasons why I wanted to write the book. First, there wasn't a good account a layman might read of the science we accomplished during Apollo. Only a few of the best known scientists had ever been recognized, and their contributions were buried either in technical journals or in stories that concentrated on the astronauts. And there was no recognition of the many contractors who assured that the science payloads would operate successfully. During my ten years of involvement in all aspects of Apollo science I had maintained a careful record of memos and reports of what was done, why it was done, who was involved and what we learned. My challenge was to put it all together in readable form.

For this book I decided that writing in long-hand as I had for the solar history book would be too difficult. Besides, I wouldn't have Chuck's secretary to type it, and I wasn't a typist. My first experience with a computer had been at BDM in Houston. We had two Apple IIs, and I would go back to the office at night to learn how to use them. I didn't want my staff to know I was a novice. Ten years later I took the plunge and bought my first computer, a desk-top Macintosh Classic, for about $1600.

I selected a Macintosh because I was told it was easy to learn how to use. That proved correct, although I never became fully knowledgeable of how to utilize all its power. I used it almost exclusively as a word processor and hadn't connected to the internet. After moving to Florida and with the guidance of my friend Roger Van Ghent I graduated to the latest iMac, the "Bondi Blue." I connected to the internet and was in business to write my second book.

I contacted Larry Cohen, MIT Press editor-in-chief of the ten-volume series on solar energy, and discussed what I was planning. Would MIT Press be interested in publishing a book that told how we selected the experiments to be carried on the Apollo missions, trained the astronauts, and the results of our efforts? Larry felt that MIT Press was not the right publisher and suggested that Bob Brugger at The Johns Hopkins University Press (JHUP) might be interested.

Using Cohen's introduction I called Brugger, JHUP's history editor, and described the story I intended to write. My timing couldn't have been better: JHUP had begun a New Series in NASA History under the direction of NASA historian Roger Launius. Six histories on different aspects of NASA programs, including a biography of NASA's administrator when President Kennedy announced the Apollo program, James E. Webb, had been published. Mine, if accepted, would be number seven. From that point I had all the incentive needed to write the book.

I had saved many filing boxes of material collected during my NASA Apollo days in anticipation of one day writing the story, but as I began writing I could see it wasn't sufficient. I needed specific sources not in my boxes to assure that the story I was telling was accurate. I called old colleagues to see if they had kept records and at first was not successful. Persistence paid off. Bob Fudali, who was on the staff of a key NASA support contractor during the early days of Apollo, Bellcomm, had saved material of events that occurred before and after I joined NASA and sent it to me. It was an invaluable record of events when the program started. Bob's records reminded me that the important role Bellcomm played throughout Apollo had never been told. I convinced colleagues that I worked with at Bellcomm that their story should be told, and that history may now be written.

Paul Lowman also became an invaluable source of material and resource for clarifying events. Paul is renowned among his NASA colleagues as a pack rat of the first degree. His office at the Goddard Space Flight Center is so crammed with reports and other memorabilia that a visitor finds it difficult to locate him among the clutter.

As I wrote sections of the book I sent drafts to Gordon Swann for review. A friend and former colleague, he participated in the struggle to develop science payloads for Apollo and was a principal investigator for two missions responsible for the astronauts' geological investigations. He provided many important comments and suggestions and a few of his famous anecdotes, some printable, some not.

I called Bob Seamans, my former boss and now a professor emeritus at MIT, to discuss a few points he made in his memoir *Aiming at Targets* that included his time as NASA Associate Administrator. During one of our conversations I mentioned that I was looking for a home for all my NASA records and, although the Smithsonian Air and Space Museum had expressed an interest, I wasn't convinced it would be the best archive. He told me a story that made me want to cry, and I don't cry very easily.

Bob had also tried to find a home for his papers and finally decided to send them to the Air Force. He boxed and sent them, I believe, to Wright-Patterson AFB. He later learned that they arrived on a Saturday and were placed in a hallway. The duty officer on his rounds asked why the boxes were there. Not receiving a good answer, he had them removed and put in a dumpster. All Seamans' extremely valuable collection, accumulated during many years while holding key positions in and out of government, was lost forever. I vowed that my collection, certainly not as valuable as Seamans' but containing some unique items, would be received with care by a willing institution.

Jim Downey, who managed my contracts at Marshall as well as others, reviewed drafts at various stages on the way toward the final product. Most importantly Jim spent many hours reviewing the files at the Marshall Space Flight Center library to select and forward to me material relating to the early years of our work. NASA's History Office was also very helpful.

Brugger sent my final draft out for reviews. They must have been good, as JHUP decided to publish and selected Alice Bennett at the University of Chicago as editor. We established a productive telephone relationship, and Alice polished the manuscript for publication. Bob Brugger, my editor at JHUP, patiently guided me through the publishing process. There is no substitute for an unflappable editor. A good selection of photographs and charts was selected, and *Taking Science to the Moon: Lunar Experiments and the Apollo Program* was released in 2001 in hard cover and then in paperback. It is now available on Kindle.

This time all reviews that I saw were favorable. Several called this a "must read" for those interested in lunar science and the Apollo program. Perhaps the most flattering stated: "Transports the reader behind NASA's facade and into the 1960s' politics, planning sessions, turf battles, camaraderie, and jealousies of the world's major space agency. An absorbing, insightful, and revealing critical history of what eventually turned out to be a hugely successful scientific endeavor." Yes! That is exactly what I had in mind. One reviewer said he hoped I would write another book. He must have been reading my mind, for as soon as *Taking Science to the Moon* was released I started writing again.

A final footnote to this story: I found an excellent home for all my NASA records. Rob Godwin, who published my next book, convinced me he would carefully archive all I would send him, and off they went. Not only have they found a good home but Rob is planning to digitize many of my records and make them available to a large audience at a new website. Thank you, Rob!

Tom Retires from Air Force - Begins Career as Southwest Airlines Pilot

After 20 years and six months of service, LTC Tom "Duke" Beattie, in 2001, retired from the Air Force. The ceremony took place at his last duty station, Spangdahlem AFB, Germany, and a short pamphlet recording his accomplishments was distributed. It included all his assignments from his student days at Lackland AFB to his retirement. He had received many commendations and decorations while logging over 3,400 flight hours, a career of which he is justly proud.

Returning to Phoenix, where he and Mary Beth had purchased a lovely home when he was an instructor at Luke, he began to search for a new career. Some Air Force pilots he had known were now working for Southwest Airlines, and they recommended that he apply. During the economic downturn of 2001 application lists for pilots were long, but Tom was hired. He now flies out of Phoenix for the only airline that continues to turn a profit year in and year out.

Terrorist Attacks of 9/11/01

We all remember important events that occurred in our lifetime and what we were doing at the time, hence the story of what I was doing when we heard the first announcement of the Japanese attack on Pearl Harbor. The terrorist attacks of September 11, 2001, in New York and Washington coincided with a trip I made to New York and Washington.

The antiwar demonstrations at Columbia in 1968 and the trashing of the NROTC armory caused Columbia's NROTC alumni to hold in contempt Columbia University administrations that kowtowed to the demands of the demonstrators. Thus, when my Columbia College class of 1951 held its 50th reunion in 2001, our NROTC contingent organized a "rump" reunion. It was held at the same time and place as the rest of the class's reunion from September 6 to 8 in Arden, New York. I didn't attend any of the official Class of '51 affairs as we had our own meeting rooms and events and held our big dinner in a separate room.

I started back from the reunion on September 9, planning on meeting Bob Brugger the next day to discuss the space station book I had just started. I wasn't able to meet Bob and spent the night of the tenth in Arlington, Virginia, where I usually stay when attending functions in Washington. The next morning I drove south on I-95 to return home. By the time I reached Richmond I had heard on a local station that a plane had crashed into the Pentagon. There was some confusion in the report, mixing the Pentagon crash with what was happening in New York and Pennsylvania. Finally it became clear that two of the planes involved had originated in Boston.

Now I was really concerned. I knew that my sister Marge and Turk were scheduled to leave Boston that morning on a flight that would connect in Los Angeles with another to China. I called Ann and suggested she ask my sister's daughters if they had any information. Calling as I drove along, Ann reported they didn't have their parents' flight number and didn't know if they were on either plane. By the time I arrived home that evening the situation was cleared up. Marge and Turk weren't on either flight, but all US domestic flights were grounded and they landed in Chicago. That was the good news, but because of the many canceled flights over the next few days and new security measures they never went on to Los Angeles or to China.

Mother's Move From Florida

Before my mother's move to Brookings, South Dakota, after my father's death she continued living alone in Bay Pines. She traveled occasionally to see my brother and sister and visited us in Maryland. She had a bracelet made with a charm for each of her grandchildren, eleven in all, and sent birthday cards to each one every year. But she had a few health issues, including diabetes. She had some problems taking her insulin injections, and Ann showed her an almost painless way to do it. But we were concerned about her living alone, and in 1991 my brother convinced her to take a small apartment at the Park Place assisted living facility in Brookings. She spent summers there and returned to Florida in the winter, insisting that she was able to live alone. On her 90th birthday in 1992, before she returned to Florida, we had a big family reunion in Brookings that included most of the grandchildren and great grandchildren.

Back in Florida she fainted and fell several times when she forgot her medications, and in 1995, returning from a shopping trip, she crashed her 1967 Plymouth into a tree. I stayed with her for a week when she was recovering from some surgery, and she baked one of my favorite desserts, as she always did when I visited, her famous lemon meringue pie. Despite her protests, we knew she could no longer stay in Florida. She finally agreed to move full-time to Brookings.

I drove to Florida to help her move and packed a few things I wanted to keep, including the song book mentioned earlier, my grade school and high school memorabilia, and a large hall mirror I had watched my father frame in Norwood. It hung in our entrance hall there and in all their homes and now hangs in our entry. After putting her on a plane I left; my brother drove down later with his pickup and met my sister. She selected a few items she wanted. My brother cleaned up the house and put it up for sale. Then he packed a few special pieces of furniture my mother wanted to have in her Brookings apartment and gave what remained to Goodwill.

And so began the final chapter in her life. Her Brookings assisted living apartment was very comfortable. It included a bedroom, sitting room with a small dining area, a balcony overlooking a park, and a small kitchen. She prepared most of her meals but could join the other residents for meals and social events as she desired. My brother and his wife Pat visited her every day, bringing the daily newspaper and doing the grocery shopping. Pat did her laundry; my mother didn't trust the one at the facility. However, every week she had her hair done at the facility beauty parlor. The parish priest visited regularly, and she established a close relationship with him. My sister Marge's family and Ann and I visited as often as possible.

In 1999 my mother moved to Brookview Manor, a full-service nursing home complex. There she stayed until her death, up to the last few days a bright, feisty, proud woman. She died on March 20, 2002, seven months shy of her hundredth birthday. According to her wishes she was cremated, and the funeral service was delayed until June so that as many family members as possible could attend. Then, again according to her wishes, I arranged to have her buried next to my father at Barrancas National Cemetery. Her name was added to his headstone. Her burial was attended by most of the immediate family, and in celebration of her life my sister and I hosted a dinner at a lovely Pensacola restaurant. It was a happy occasion, with few tears. We all told stories, mostly humorous, of how she had affected our lives.

VS-30 Reunions

My old squadron, VS-30, held reunions every two years starting in the fall of 1993 in locations around the country where former squadron mates had scattered. One or more of the guys would volunteer to organize and host each reunion. Ann and I didn't make the first one in Pensacola but thereafter were able to attend many, including one in Monterey, California, in 1997. Among highlights of that reunion we met Mary Beech, my fellow high school Dodger fan, for lunch and caught up on her California life and that of other classmates living there.

I volunteered to host the 2003 reunion in Jacksonville. It was a good location. The VS-30 active duty squadron was stationed at the Jacksonville Naval Air Station and the aircraft carrier John F. Kennedy was home-ported at the Mayport Navy Base just east of the city. With Jackie Paulson, the widow of a former VS-30 commanding officer living in Ponte Vedra, we arranged accommodations, a ready-room in the hotel, tasty hors d'oeuvres, and three full days of activities including a guided tour of the Kennedy. We were invited to a reception held in the squadron's ready-room at the airfield. It had been many years since most of us last sat in a ready-room, and some of the changes were surprising.

Among others, it contained a bar, and the squadron's two women pilots attended. Women pilots! What sort of nonsense was that? (I admit to being a chauvinist, but hurray for women! What would we do without them?) Our reunion ended with the traditional sit-down dinner at the Orange Park Greyhound Kennel Club, where a special race was dedicated to the squadron. Our guest speaker, the VS-30 commanding officer at the time, CDR James Gregorski, filled us in on the squadron's recent operations and life in the modern Navy.

The 50th anniversary of VS-30's commissioning as an active duty squadron during the Korean War was celebrated in December 2003. Ann and I attended the ceremony that highlighted its many accomplishments. A former squadron mate, Paul Canada, and I were the only attendees who had been in the squadron in 1953 when it transitioned from a reserve squadron to active duty. I was asked to tell a few tall tales, and my home movies of carrier landings in 1953 and 1954 were shown continuously on a large TV in the ready-room. My reward for telling 50-year-old stories was a handsome framed photograph and plaque showing two AFs in flight, the plane flown by the squadron when it was first commissioned.

Two years later we participated in a sad ceremony, VS-30's decommissioning. One of the last VS squadrons still active, it had recently returned from a deployment supporting the war in Iraq. By now it was flying the S-3 Viking, a two-engine jet made by Lockheed. One of the squadron's primary duties during the deployment was using the S-3 as a tanker. After takeoff from the carrier with a full armament load, the F-18 Hornets and EA-6 Prowlers rendezvoused with a Viking and topped off their fuel tanks before departing for the war zone. This valuable support activity provided a little safety margin by reducing their takeoff weight. The squadron returned to Jacksonville knowing that it had been their last deployment under the logo of a black panther reclining on a white cloud. VS-30 is now a part of Naval aviation history, and I am a proud alumnus.

Superbowl XXXIX

Superbowl XXXIX was held in Jacksonville in 2005. Along with thousands of others I volunteered to help make it a success and raise the profile of the Jacksonville area in the sport and business universe. Many of the volunteers from St. Johns County joined to make the VIPs who attended aware of opportunities to do or locate their business in the county.

Training sessions for the many different volunteer jobs were held in Jacksonville. I joined Len Weeks, former St. Augustine mayor, and the county Chamber of Commerce to develop a program promoting St. Johns County. We knew that many VIPs would fly into the St. Augustine-St. Johns County airport, located just north of the city, where their planes would be parked on an off-duty runway until they left. The arriving passengers would be a captive audience for whatever programs we could put together.

We decided on a two-pronged promotion. We would greet the arrivals at the entrance to the airport terminal with a package containing information about the county and locally made treats such as chocolates, spicy sauces and potato chips. Inside the terminal we would have an information booth and actors in period costumes reflecting the history of St. Augustine as the country's oldest city. On their departure we would have a different package with more treats to be consumed on the way home and information on the county's business community.

We assembled hundreds of bags; we knew approximately how many planes and passengers would be arriving by the flight plans they had filed. We manned the airport from early in the morning until late night for two days before the game and one day after to make sure we greeted almost everyone. A few of the VIPs were sneaky; their limos met them at their planes after they parked and drove off without coming to the terminal. I was one of the greeters; my official uniform of bright red hat and wind breaker worn for a few hours still hangs in the closet.

Did we succeed in publicizing St. Johns County? That's debatable. However, without our hotels and other county facilities Superbowl XXXIX would not have run so smoothly. Both teams, the Philadelphia Eagles and New England Patriots, and their entourages stayed in St. Johns County hotels. The Patriots practiced at our new Bartram Trail High School fields, only a ten-minute bus ride (with accompanying police escort) from their hotel at World Golf Village.

Third Book

Having been involved in 1982 during the early NASA planning to build a space station, and later serving seven years on the NASA Space Station Advisory Committee, I believed I was in position to write a well documented history of the program. When I began writing in 2001 the space station, now called the International Space Station (ISS), was facing many problems. Perhaps describing how we had arrived at the present position would be useful to decision makers as they debated how to keep the program moving ahead.

The history I decided to write would cover the development of the space station from 1982 to the launch of the first two elements in 1998.

I would provide brief annual summaries carrying the program from that point until submission of the manuscript for publication. I would describe how the program struggled to survive in a political environment that constantly modified its goals and content. I would also describe the effects of powerful antagonists who tried to have the program terminated. In the final chapter I planned to summarize the failures of Congress, several administrations and NASA management to spend taxpayer funds wisely.

The book would document a space station placed in orbit six years later than originally projected, far over budget, operating in an impaired and incomplete condition. Before starting I had determined that a complete space station history would require a book of enormous length. Therefore, I chose only the details, people and events that would best help the reader understand the climate in which NASA managers operated. At the time I began writing my story the Space Station faced a very uncertain future.

I sent early drafts to Bob Brugger, hoping he would agree to publish the book. He sent them to several reviewers. Their comments were not very encouraging. One actually said it wasn't worth publishing. Undeterred, I continued writing and contacting other possible publishers. Getting a book published is a tough business, even for a successful author.

By chance I talked to Jack Schmitt, who suggested contacting Apogee Books, a publisher in Canada that specialized in books on space topics. He was familiar with Apogee and gave me the name of the president, Robert Godwin. I called Godwin and told him that I had an almost complete manuscript. He was interested, especially when I said I wasn't asking for any financial support. The result has been a very productive and friendly relationship that continued after Apogee published my book.

"Who Was in Charge? No One!" is the title of the final chapter, which was the most difficult to write. In a few pages I summarized all that went on in the 18 years before the first elements of the ISS reached orbit. Actually I developed an outline for this chapter and wrote part of it before writing about how the program started. It had to encapsulate all the problems and concerns that I believed shaped the program. When I began writing the rest of the story I often returned to this chapter to assure it was complete and, most importantly, included lessons learned. It ends with a forecast of NASA's future and the problems I believed NASA would face.

The book's title, *ISScapades: The Crippling of America's Space Program,*" provides a preview of what would be found between the covers. It is a takeoff on the ice show and is an appropriate metaphor for a program that skated on the brink of disaster for many years. From the time space station planning began in 1982 until the first elements were operating in orbit in 1998 some 220 miles overhead, the space station impacted all the nation's civil space programs. The constant cost overruns required NASA and Congress to raid the budgets of other NASA programs to keep the space station moving. Promises were always made to restore the raids, but Congress never saw fit to provide the additional funds. The finished book, published in 2006, included detailed discussions of budget issues. In the last analysis funding, or the lack thereof, shaped the program as much as or more than anything else.

ISScapades received many complimentary reviews. It was called a "tour de force" in one, "not a fluff NASA history" in another. Jim Oberg, a well known author and space commentator and former NASA engineer, complained that page margins were too narrow for notes, but wrote in the IEEE magazine *Spectrum*: "Would-be space managers should have to read and digest this book to prove that they, like the astronauts, have the 'right stuff'."

With *ISScapades* in book stores I had exhausted those subjects about which I could write authoritatively at book length. I turned to writing articles for various professional journals and trade magazines and the occasional letter to the editor to explain why a NASA program or energy R&D was proceeding in a wrong direction. Then, in 2006, I began this long journey from the 20th to the 21st century.

Afterword

There are a few loose ends to tie together. Remember the cameo I bought in Naples in 1951 for my mother? I never gave it to her; why, I don't remember. It reappeared 55 years later, still in the original box. Perhaps the tooth fairy had it hidden away. I had it made into a brooch on a long gold chain and presented it to Ann on her birthday. Ann found few occasions to wear it and gave it to Mary Beth as a birthday present. Thus, hopefully, this "family heirloom" will never be lost again and future wearers will know its special history.

Remember the story about Phoebe Gilmore and her cookbook? After working for several years on the manuscript, she thought she had found a publisher. She sent her only copy and never heard back despite many efforts to contact the company. Anyone purchasing a cookbook featuring microwaveable Mexican dishes may be seeing Phoebe's efforts. She died of emphysema a few years ago, so she will never know if someone is enjoying one of her recipes. We stay in touch with Bill and he has visited us several times. He now lives in Hammondsport, New York, and spends his free time as a docent at the Curtiss Museum. Among other activities he has helped build flying replicas of Curtiss' airplanes, the Navy's first purchase, the A-1 Triad, and others.

I am confident that this history would be much more complete if I had started it, as I always intended, while my mother was alive. To the very end, in her ninety-ninth year, she had a remarkable memory that would have allowed me to verify and add to my stories. Readers, procrastination will always catch you in the end!

As I finish this story the grandsons, now young men of 22 and 19, have begun their life adventures. Travis, after a year at Arizona State University, suddenly decided to enlist in the Marines, perhaps not so suddenly as years ago he had expressed admiration for those who serve in the Special Forces of our military. He finished boot camp with high honors and refused to be considered for the Marine detachment in Washington that provides ceremonial services. He wanted to go to a combat unit. Attached to the Marine 7th Regiment, 2nd Battalion, Fox Company, he served in 2008 in Helmand Province, Afghanistan, one of the most dangerous areas. He participated in many engagements against the Taliban and was promoted to lance corporal. Although injured by an IED explosion, when he returned to the States he decided to re-enlist. Whatever he chooses to do in the future, a proud grandpa believes the sky is the limit.

Kyle has followed a different path. When Tom was stationed in Germany Kyle and Travis were selected to play on a German ice hockey team and progressed to be excellent players. Three years ago Kyle played on the US Under 17 team at an international tournament in Czechoslovakia. He is now playing hockey on a scholarship at the University of Maine. If he is good enough and it is what he wants to do, he will play hockey professionally. If not, he is a very bright young man, taking a full academic schedule at Maine majoring in business management. I know he will be successful in his next steps whatever they may be.

Ann and I have settled into a comfortable routine. At breakfast we sit quietly, each reading a newspaper (we subscribe to two daily papers and the *New York Times* on Sunday so Ann can read the book reviews), then go on to daily chores maintaining the house and gardens. Ann continues to cook great dinners for friends and family, although when eating alone we often enjoy carryout meals from our favorite local restaurants.

While Ann continues with household chores normal to wives of our generation, and after finishing my assigned chores such as taking out the garbage, I am free to spend time reading three more newspapers online and numerous journals sent by professional organizations to which I belong. Then I write the occasional article or prepare the occasional lecture to be delivered in Washington or some other city.

Recently I was invited to be the keynote speaker at an international conference on drilling to the Mohorovicic discontinuity, the boundary between the Earth's crust and mantle, a geological mystery zone. I was invited to Huntsville, Alabama, for the 40th anniversary of the first Apollo mission (Apollo-15) that used the Lunar Rover built under the direction of the Marshall Space Flight Center. And in 2011 I will attend a few ceremonies celebrating the 100th anniversary of Naval Aviation. If you live long enough you are invited to attend many anniversaries and ceremonies that celebrate events of long ago. But that doesn't happen very often, a far cry from the days of frequently rushing to airports to catch a long-distance flight or waking up in a rain forest camp on a misty morning to explore a forested canyon or wade a quiet-running stream.

That former active life, including being bucked off cantankerous mules (they do buck like horses when startled) and years of participating in sports, have taken their toll on these old bones. Two hip replacements and a bum knee make airport security irritating and time-consuming. My long legs rebel against being confined to a narrow, hard aircraft seat, even for a few hours, so I drive to Washington or other destinations. It takes longer but is more convenient and provides time to contemplate the problems of a complex and dangerous world. My ability to help find solutions is now greatly diminished. However, I never give up trying, much to Ann's despair.

I feel very fortunate to have lived in a unique country on this small planet where the only limits to my being successful were self-imposed. I am unashamedly a flag waver even though, at times, I strongly disagree with policies or directions our elected leaders have taken. It often appears they have forgotten or ignored hard-won history lessons. My recommendation: before beginning a new venture in foreign affairs all our leaders should refer to Winston Churchill's six-volume history of WW II. If they don't own a set, they should buy one.

As I observe the world around me I am deeply troubled by what I see. It seems to be very different from the one in which I grew up. I am not alone in believing that many in all societies have embraced a life of self-indulgence and a goal of gaining fleeting notoriety, encouraged by new and easy methods of communication that focus on the lives of the great and near-great.

Some historians claim that those born in the first half of the 20th century grew up in a more innocent time. I disagree. How could anyone consider it a period of innocence? There were ruthless political upheavals, a worldwide depression and two world wars during which millions of true innocents died. Hopefully, future historians will record that those who survived the turmoil set the stage to build a better world for all. Idealistic? Yes, although one writer has already labeled those who brought World War II to a successful conclusion as the "greatest generation." Will their descendants complete the needed changes? It won't be easy.

With the above in mind, I am reminded of an old Noel Coward song: *What's going to happen to the tots?* In case you are not familiar with the lyrics they go like this:

"Life today is hectic,

our world is running away.

Only the wise can recognize

the process of decay.

"What's going to happen to the children

when there aren't any grown ups around

to help them find their way?"

The last two lines above are courtesy of the author with apologies to Mr. Coward. However, the remaining original lyrics, a total of 72 lines, are as relevant today as when first written in 1955.

Final Thoughts - Pondering the Imponderable

The manner and time of one's death is being not predictable, a fall in the shower, being run down by a drunken driver, a brain tumor. It is interesting that the longer one lives the longer actuarial tables predict he will live. Logically, if such tables are to be believed, one should live forever. But eventually, no matter how strong one is or how carefully he has lived his life and eaten his vegetables as mother commanded, the body decides it will no longer repair the constant wear and tear and shuts down. The end may come gracefully or not, and probably not of one's choosing.

Most men believe they are indestructible and invincible and find it very hard to admit otherwise. However, nature appears to have prepared women better for the inevitable. Dealing with serious sickness or death of a child or loved one usually becomes the duty of women. Perhaps it is easier for them because they know how to cry unashamedly. Probably learned scholars have analyzed this aspect of human nature and revealed the reasons. If so, I have yet to read their conclusions.

And so, as I approach the final days of existence as a cognitive being, the question naturally arises - what's next? And perhaps a more frightening question is there a next? For those with strong religious beliefs this doesn't seem to be a problem; they will be rewarded or punished depending upon how they lived. However, as an agnostic I have arrived at a logical conclusion that calms my fears and answers my questions. It is based on the realization that I have no recall of my existence prior to a few years after my birth. What was I doing, for example, 200 years ago? Where was I?

The answer is, of course, that I didn't exist, at least not to my recollection. All the needed DNA from my father and mother hadn't combined until sometime in February 1929. And the DNA provided by their parents, which eventually was included in mine, hadn't combined some 30 years earlier. Back and back one might go if he subscribes to evolution. So, perhaps I existed in some molecular form from the beginning of the human species, modified slightly in every generation over the millennia. And before that, from the time life first began on the planet, a gene or two survived in each succeeding generation until the first creatures slithered out of the sea to walk on dry land and eventually homo sapiens appeared.

Funeral ceremonies often end with the ageless saying, "We commit this body to the ground, earth to earth, ashes to ashes, dust to dust." However long ago this became a final farewell, it reflected the knowledge and beliefs of the time. One's earthly being ended as a few grains of dust, but family traits of those still living continued in some mysterious way. Today we know that those who have created children live on in their genes and those of their children's children.

Would you consider that comforting? I find it so and can only hope that I have passed on genes that will be of value to future generations.

Our Florida home

Bill Gilmore with Curtiss' America he helped build, Hammondsport, New York

Phoebe Gilmore and Ann on a Puerto Vallarta, Mexico, vacation

Bob Kovach and Jack Schmitt at 30th anniversary of Apollo-17 mission

With Tom and Pat at Crescent Beach

Family at Mother's burial with Dad at Barrancas National Cemetery, Pensacola

VS-30 2001 reunion - Bob and Maggie Timm at left,
Wit and Carol Johnson at right

Hosting the 2003 reunion in Jacksonville

Visiting the USS John F. Kennedy at Mayport Naval Base during reunion

Attending the 50th anniversary of VS-30's active duty at NAS Jacksonville

Travis, Kyle, Mary Beth and Tom in backyard, Buck Eye AZ 2011

Space Station configuration in 2006 when ISScapades went to press

Astronauts Charlie Duke and Jack Schmitt at 40th anniversary of Apollo LRV

Rob Godwin hard at work in his Ontario, Canada, office

Roger Van Ghent - Apple/Macintosh guru

Index

Acandí, Colombia	167, 171
Adair, Andrew - Chicago Daily News	10
AD Skyraider - Navy airplane	6, 134
AEC - Atomic Energy Commission	243, 249, 250
AF Guardian - Navy airplane	6, 115, 122, 123, 140
AID - US Agency for International Development	8, 259, 261, 262, 279, 293, 294, 295, 296, 297,
Aldrin, Edwin "Buzz" - NASA	205, 227, 238
Allenby, Richard - NASA	181, 200, 215
Alpine, NJ	34, 41
ALSEP - Apollo experiment package	223, 225, 226, 231
Andrews, Edward - NASA	182, 196, 197, 207
Annapolis, US Naval Academy	67, 68, 70, 72, 77, 83, 94
Antietam, USS - CVA 36	136, 137
Apogee - Book publisher	1, 234, 330,
Apollo - NASA moon program	7, 180, 181, 182, 196, 197, 198, 199, 200, 201, 202, 203, 204, 205, 206, 207, 208, 209, 210, 211, 212, 214, 215, 216, 218, 219, 220, 221, 222, 223, 224, 225, 226, 227, 228, 229, 230, 231, 232, 233, 234, 235, 236, 237, 250, 254, 258, 269, 270, 271, 272, 289, 291, 292, 296, 299, 322, 323, 324, 333, 337, 342
Archaeology	65
Argyrol - medicine	17
Armstrong, Neil - NASA	205, 227, 238, 270
Ashkhabad, Turkmenistan	246
ASW - Anti-submarine warfare	6, 114, 124, 134, 136, 146
Athens, Greece	94
Atrato River, Colombia	165, 166, 167, 189
Bailieborough, Ireland	133
Baker Field, Columbia University	73, 75, 82
Baku, Azerbaijan	246, 247
Balboa, Vasco Nuñez de	102, 163, 164, 170
Barber, Red - sports announcer	33
Barcelona, Spain	128
Barin Field, FL	6, 112, 116, 120
Barr & Wright - 1870 photographers	283, 297
Bartlesville, OK	282
Bartram, William - 18th century naturalist	318, 329
BDM - Company	8, 265, 280, 281, 282, 283, 284, 322
Bear Mountain, NY	28, 48, 49, 75
Beaufort, USS - AK-6	16, 17, 55
Beech, Mary	4, 44, 62, 327
Beeville, TX	146
Beggs, James - NASA	264
Belfast, Ireland	131, 132, 133
Bellcomm - NASA Apollo contractor	323
Bever, Jim - AID	294, 295
Bilateral - Agreements between nations	7, 244
Blue Angels - Navy flight team	67, 111
Boeing - Company	8, 182, 200, 289, 290, 291
Brookings, SD	314, 326
Brooklyn Dodgers	33, 35, 44, 80
Brugger, Bob - JHUP editor	322, 324, 325, 330
Bush, President George H. W.	8, 255, 288, 289, 310
Camp Shanks, NY	45, 46
Canal Zone, Panama	102
Cannes, France	90, 96
Cartagena, Colombia	171, 179, 194, 263, 285
Carter, President James Earl, Jr.	31, 252, 253, 254, 257, 262
Casanare River, Colombia	172, 173, 192
Cernan, Eugene - NASA	231, 232, 233, 272

344

Chattanooga, USS - CL-16	16
Chevron - Oil company	281, 284, 285, 287, 305
Chicago, IL	10, 13, 26, 28, 35, 40, 56, 224, 242, 278, 324, 326
Chocó - Indian tribe	164
Cicuco - Colombian oil field	172
Closter, NJ	22, 40, 41
Cold War	93, 129
Collins, Mike - NASA	227
Colombia, South America	4, 6, 7, 8, 143, 147, 149, 150, 151, 152, 153, 154, 155, 156, 157, 162, 163, 165, 166, 167, 171, 172, 173, 175, 176, 177, 178, 179, 180, 181, 182, 185, 212, 214, 216, 228, 229, 263, 268, 284, 285, 286, 287
COLPET - Mobil/Texaco company	154, 175, 178, 180
Columbia - College/University	5, 44, 50, 51, 63, 64, 65, 67, 74, 75, 76, 77, 80, 81, 82, 83, 84, 85, 88, 89, 91, 103, 110, 112, 133, 135, 138, 145, 224, 259, 325
Columbus, USS - CA-74	67, 72, 117
Connecticut	28, 34, 35, 37, 130, 138, 195
Copernicus - Moon crater	206
Corpus Christi, TX	6, 114, 116
Corry Field, FL	6, 109, 111, 116
Costa Rica	8, 283, 284
Cougar - Navy jet	146
Coveñas - Colombia oil terminal	154, 178, 286
Cresskill, NJ	41, 44
CSM - Apollo spacecraft	201, 202, 205, 206
Cubs - Baseball team	33, 35, 40
Cuna - Indian tribe	164, 165, 170
Curtiss, Eugene - Airplane pioneer	78, 87, 332, 336
Cuzco, Peru	212, 213, 214
Darién - Colombia province	6, 163, 164, 165, 168, 170
Dekker, Larry - Chevron	285
De Laurier, James - Univ. of Toronto	300, 301
Desert Storm	298, 300
Dilbert Dunker	106, 119
Downey, James - NASA	208, 209, 324
Dublin, Ireland	248
EASEP - Apollo-11 experiment package	225, 226
Eason, Bill - USS Liddle	91, 101, 103, 104
Eggers, Alfred - NSF	236, 243, 244, 249
Eisenhower, President Dwight D.	83, 288
El Caribe - Cartagena hotel, Colombia	179, 263
Erie Railroad	23, 41
Essig, Henry - Norwood	22, 36
Evans, Thomas C. - NASA	181, 182, 196, 197, 200, 201, 203, 207, 233
Evinrude - Outboard motor	47, 48
FCLP - field carrier landing practice	112, 113, 116, 125, 136
Fentress Field, VA	123
Floresanto - Colombian oil field	286
FM - Radio	34
Ford, President Gerald R. Jr.	245, 249, 253, 257
Fordyce, Sam - NASA	299
Foster, Willis - NASA	200, 202, 207, 223, 237
FRC - Fuel Research Center, Pakistan	296, 297
Frosch, Robert - NASA	260, 262, 263
Frosco, Jimmy - Norwood	38, 41, 43, 44, 46, 60, 112
Fryklund, Verne - NASA	197, 198, 199, 200, 202
Garrard, Frank - VS-30	105, 106, 114, 117, 121, 128, 130, 136
GE - General Electric	290, 297, 298, 299
Gemini - Space program	7, 202, 207, 211, 212, 214, 216, 218, 221
George Washington Bridge	18, 45
Germany	20, 27, 37, 43, 45, 52, 157, 168, 178, 241, 242, 298, 299,

	304, 321, 325, 332
Geysers - Geothermal power field	244
Gibraltar	71, 72, 73, 128, 139
Gillespie, Rollin - NASA	212
Gilmore, Bill/Phoebe	176, 177, 179, 281, 331, 336, 337
Gimbels - Department store	26
Glenn, John - NASA	180, 251
Goddard Space Flight Center, MD	181, 195, 198, 208, 323
Godwin, Robert - Apogee publisher	4, 234, 324, 330, 342
Golden, CO	44, 138, 143, 148, 151, 317
Gravetye Hotel, England	247
Gray, Edward "EZ" - NASA	200, 203, 204
Great Depression	1, 5, 10, 13, 14
Haffer, Jürgen/Maria - Mobil Oil	4, 152
Haliburton - Oil well company	174
Hampton Roads, VA	5, 80, 99, 100
Hedburg, Hollis - Chevron	306
Herwig, Lloyd - NSF	248
Hess, Harry - NRC	220, 229
Hinners, Noel - NASA	292
Hirsch, Robert - ERDA/DOE	250, 252
Hollingshead, Charles - NASA	234
Houston, TX	8, 199, 208, 215, 220, 227, 232, 235, 238, 239, 240, 252, 265, 280, 281, 282, 283, 293, 302, 322
Hughes - Company	298, 299
Hunt, Dave - VS-30	137, 138
Illinois Institute of Technology	16
Instituto de Ciéncias Naturales, Colombia	169
Iona Island, NY	49
Ireland	52, 131, 132, 133, 240, 248, 303, 304
Irwin, James - NASA	271
Islamabad, Pakistan	294, 295
jaguar	168
Johnson, Harry - ERDA	4, 19, 203, 260, 265, 266, 282, 283, 288, 291, 293, 339
JPL - Jet Propulsion Laboratory	230, 259, 262, 264, 265, 300
Julington Creek Plantation, FL	316
Kamchatka,USSR	244
Karachi, Pakistan	294, 295, 296
Keane - Ann's family	238, 276
Kelly, Bill - Columbia	80, 84, 88, 89
Kepler - Moon crater	206
Kingsville, TX	6, 114, 115, 116, 117, 121, 146
Kohlberger, Johnny - Norwood	38, 41
Korean War	52, 144, 328
Kowalski, Bob - VS-30	4, 121, 128, 130, 131, 133, 134, 137, 140
Kraft, Christopher - NASA	215, 260, 272
Laguna de Guatavita - Colombia S.A.	268
La Heliera-1 - Colombian oil well	172, 174, 175
Lahore, Pakistan	294
La Rada-1 - Colombian oil well	171, 172, 173
Leghorn, Italy	71, 72
Leopold - ex-King of Belgium	164, 170
LERC Lewis Research Center	251
Liddell, Urner - NASA	181
Liddle, USS - APD-60	5, 89, 90, 91, 92, 93, 94, 95, 97, 98, 102, 103, 104, 106
Link - Instrument flying trainer	109, 110, 116, 117
Lion's Den - Columbia College	74, 82, 84
Lisbon, Portugal	69, 71, 304
Little Creek, VA	5, 76, 79, 80, 89, 93, 97, 98, 99, 100, 101
Llanos, Colombian eastern plains	172, 173, 175
LM - Apollo lunar module	201, 204, 205, 206, 225, 226, 227, 268
Lobeck, Armin K. - Columbia	65

Loewenthal, Rudolf/Ariadne - Neighbors	241
Lombardi, Vince - Coach	44
Lovelace, Al - NASA	257, 258, 259, 260, 262
Lowman, Paul D. - NASA	198, 202, 205, 211, 323
LRV - Lunar roving vehicle	231, 232, 342
LSO - Landing signal officer	112, 113, 123, 124, 126, 128, 140
Luck, Marie - Norwood	27, 30, 60
Lukjanow, Wladimir B. - Maryland	241
Lunokhod - Soviet robotic lunar rover	222
Macon, USS - CA-132	67, 68, 69, 70, 72, 87
Malkin, Myron - Consultant/Partner	290, 297, 299, 302, 308
Malta	95, 96
Marseille, France	96
Marshall Space Flight Center, AL	202, 208, 258, 288, 324, 333
Martin Marietta - Company	204, 292
McCloskey, Jhea - VS-30	130, 137
Mercury - Space program	202, 207, 211, 221
Merner, Carl - Columbia	64, 75
Meteor Crater, AZ	203, 209, 211, 267, 268
Miller, LCDR Kenneth - USS Liddle	90, 94, 95
Mindoro, USS - CVE-120	6, 97, 123, 125, 126, 128, 129, 139, 140
MIT	206, 302, 303, 317, 318, 322, 323,
Mobil Oil - Company	150, 216, 237
Mompós, Colombia	154, 172
Mondale, Vice President Walter F.	254
Monterey, USS - CVL-26	113, 116, 123, 327
Monteria, Colombia	152, 154, 155, 158, 162, 166, 171, 175, 190, 191, 263, 285, 286, 287, 305
Moscow, USSR	241, 244, 245, 246, 247, 249, 275
Motilone - Indian tribe	154
MSC - Manned Spaceflight Center, TX	199, 207, 220, 221, 225, 238
Muehlberger, William - NASA	232, 233
Mueller, George - NASA	200, 204, 206, 218, 223, 254
Muzo - Colombian emerald mine	179
Myers, Dale - NASA/DOE	254, 255, 256, 257
Naples, Italy	5, 46, 96, 97, 128, 331
NASA	7, 8, 180, 181, 182, 195, 196, 197, 198, 199, 200, 201, 202, 203, 204, 206, 207, 208, 210, 211, 212, 214, 215, 216, 217, 219, 220, 221, 222, 223, 225, 226, 227, 228, 229, 230, 232, 233, 235, 236, 237, 242, 250, 251, 252, 253, 254, 256, 257, 258, 259, 260, 261, 262, 263, 264, 265, 277, 279, 282, 283, 288, 289, 290, 291, 292, 293, 299, 300, 301, 317, 322, 323, 324, 329, 330, 331
National Space Club	282
New, USS- DD-818	83
New York Central Railroad	51
Nixon, President Richard M.	236, 243, 244, 249, 254, 288
Norfolk, VA	6, 67, 72, 73, 83, 87, 97, 100, 102, 105, 117, 122, 124, 125, 126, 128, 129, 130, 131, 133, 134, 135, 137, 138, 143
North, LTCOL Ollie - National Security Council	284
Northvale, NJ	18, 23
Norwood, NJ	4, 5, 17, 18, 19, 20, 21, 23, 24, 28, 29, 35, 36, 38, 39, 40, 41, 42, 57, 58, 59, 63, 73, 82, 85, 326
NRC - National Research Council	259, 260, 261
NROTC - Naval Reserve Officer Training Corp	5, 50, 51, 63, 64, 66, 67, 68, 72, 73, 77, 84, 89, 117, 224, 325
NSF - National Science Foundation	236, 237, 243, 244, 245, 246, 250, 251, 253, 258, 265, 301, 302
Oceana Naval Air Station, VA	123, 135
O'Leary, Jack - DOE -	254, 255, 256, 257
OMSF - Office of Manned Space Flight	197, 198, 206, 223
ORI - Operational readiness inspection	98, 99, 100, 101
OSS - Office of Space Science	181, 197, 198, 200, 208, 223
Pan American Highway	165

PBM Mariner - Navy seaplane	79
Pearl Harbor, HI	37, 325
Pensacola, FL	4, 6, 76, 77, 78, 79, 84, 100, 103, 105, 107, 109, 113, 116, 117, 121, 229, 327, 338
Peshawar, Pakistan	294
Peterson, "Mitzi" - NASA	258
Petróleos de Venezuela - Oil company	281
Phillips, Lt. Gen. Samuel - NASA	200, 223
Philly - Philadelphia, PA	75, 76, 81, 82, 83, 84
Piermont	45, 47, 48, 49, 50
Porthcawl, Wales	248
Port Lyautey, Morocco	90
Portsmouth, VA	98, 103
POWs - Prisoners of war	45
Press, Frank - NASA	302, 303, 317, 318, 322,
Quayle, Vice President James "Dan"	288, 289
Quonset Point, RI	90, 131
Ranger - Spacecraft	32, 210, 211, 214, 222
RANN - Research Applied to National Needs	236, 237, 238
Reagan, President Ronald W.	252, 255, 264, 284
Red Team	289, 290
Rekenthaler, Doug -BDM	283
Reynosa, Mexico	238, 239
RFP - Request for proposal	289, 291
RFQ - Request for quotation	207, 209
Rhoad, Merritt - NROTC	64, 73, 88
Rockville, MD	168, 195, 240, 266, 283, 289, 316
Rollins, Hazel - DOE	195, 255
Rondón, Colombia	173
Roosevelt, President Franklin Delano	37
Rubey, William - NRC	230
S2F Tracker	6, 130, 133
Sammons, LCDR Robert - USS Liddle	94, 95, 96, 98, 99, 100, 101, 102
San Jacinto Mountains, Colombia	178, 180
Saufley Field, FL	6, 110, 111
Sautatá, Colombia	167, 171
Scherer, Lee - NASA	181, 182, 223, 238, 239
Schleef, Jack - Columbia	80, 88
Schlesinger, James - DOE	253, 254, 255, 257
Schlumberger - Oil well company	174
Schmitt, Harrison "Jack" - NASA	220, 221, 231, 232, 233, 258, 272, 330, 337, 342
Scott, Dave - NASA	205, 270, 271
Seamans, Robert C. Jr. - NASA/ERDA	250, 251, 252, 261, 323, 324
Shepard, Al - NASA	219, 293
Shoemaker, Gene - USGS	196, 197, 202, 209, 217, 219, 267
Sicuani, Peru	212, 213, 214
Sinú River, Colombia	154, 155, 189, 190, 285, 286, 287
Sixth Fleet	6, 68, 89, 93, 94, 97, 98, 126, 127, 129, 131
Slayton, Deke - NASA	219
Smithsonian Institution	253
SNB - Navy airplane	116, 118
Snellen, James - Florida	92
SNJ - Navy basic trainer	79, 107, 109, 110, 115, 116, 118
Sonnett, Charles P. - NASA	198, 199, 200, 203, 206, 214, 215
Souda Bay, Crete	94, 95
Soviet(s)	7, 43, 93, 129, 131, 135, 180, 222, 241, 244, 245, 246, 247, 248, 249, 284
Spalding, Carl - Mobil Oil	150, 167, 171
SSAC - Space Station Advisory Committee	292, 293
SSPS - Space Solar Power System	260, 261
Staten Island, NY	10, 13, 14, 16, 24, 53, 54
Sticco, Bobby - Norwood	23, 41, 43, 60, 62, 75, 112, 119

Stuhlinger, Ernst - NASA	208, 209
Surveyor - Spacecraft	210, 211, 222
Swann, Gordon - USGS	210, 323
Swanson, Ed "Toad" - USS Liddle	93, 104
Sylvester, Wendell "Doc" - NROTC	64, 76, 89
T-38 trainer	220
Tacarcuna Mountains, Colombia	166, 189
Tamaulipas, Mexico	238
Tappan Zee, NY	45, 47
Tashkent, Uzbekistan	245, 247
Tbilisi, Georgia	247
TBM - Navy WW II airplane	6, 113, 114, 115, 116, 117, 118, 121, 122, 123
Teem, John - ERDA	250
Tenafly, NJ	4, 40, 41, 42, 43, 44, 45, 51, 52
Tenché-1 - Colombian oil well	172
Tequendama Hotel, Bogota, Colombia	151, 156, 263
Terwilliger, Harvey - Norwood -	4, 35, 41, 60, 62, 86
Texaco - Company	153, 154, 179
Tierra Alta, Colombia	155, 156
Timm, Bob - VS-30	131, 132, 339
Townes, Charles - NASA	225
Trilobites	175
Troub, Charlie - McAllen, TX	238, 239
Turbo, Colombia	165, 166, 167, 168, 171
Turner, Rhett - VOA reporter	227, 233
UDT - Underwater Demolition Team	90, 91, 93, 94, 95, 96, 97, 104
UFO	132
Urabá, Gulf of, - Colombia	163, 168, 285
Urich, Tom - USS Liddle	90, 91, 98, 104
US Agency for International Development	293
USGS - US Geological Survey	196, 197, 206, 209, 210, 211, 222, 231, 232, 270, 296
Valley Forge, USS - CVS-45	131, 132
vampire bats	160, 180
Van Cortlandt Park, New York City	74
Van Ghent, Roger - Florida	4, 322, 343
VE Day - Victory in Europe	45
VF-712 - Navy squadron	146
Virginia Capes	98, 99, 123, 125
Visser, Roger - VS-30	128
VJ Day - Victory in Japan	46, 83
VOA - Voice of America radio	227, 228, 233, 234, 265
von Braun, Wernher - NASA	202, 208
VS-30 - Navy squadron	4, 6, 9, 98, 117, 122, 130, 138, 139, 327, 328, 339, 340
VS-661 - Navy squadron	218
VS-751 - Navy squadron	146, 218
Walden, Omi - DOE	253, 257, 258, 262
Western Electric Company	16, 25, 37
West Norwood, NJ	23, 35, 82
West Point Military Academy	48
Westwood, NJ	31
Whiting Field, FL	6, 106, 107, 112
Winnebago - Motor home	226, 228, 229, 241, 270
Worden, Al - NASA	271
Wright Brothers	22, 78, 115, 134, 135, 138
WW I	16, 27, 28, 37, 55
WW II	13, 45, 64, 67, 68, 70, 78, 79, 84, 90, 92, 94, 105, 111, 113, 114, 123, 131, 144, 167, 208, 242, 333
Yerevan, Armenia	246
Yucatan Peninsula, Mexico	229
Yugoslavia	8, 321

References

• Pakistan Energy Planning and Development Project Evaluation Report, Submitted to AID by CBY Associates, Inc., July, 1988

• Institutional Assessment of the Fuel Research Centre of the Pakistan Council of Scientific and Industrial Research, Submitted to AID by CBY Associates, Inc., June, 1989

• History and Overview of Solar Heat Technologies, D. A. Beattie, C. Bankston, F. Morse, MIT Press, 1997

• Taking Science to the Moon, D. A. Beattie, The Johns Hopkins University Press, 2001

• ISScapades: The Crippling of America's Space Program, D. A. Beattie, Apogee Books, 2006

The author obtained NASA funding to reproduce a Lunar Orbiter photo of the Moon on a volcanic cinder field east of Flagstaff for astronaut training.